*For Linda
Enjoy* oc

# Go For It

*Volunteering
Adventures
on Roads
Less Travelled*

A memoir by

ALASTAIR HENRY and CANDAS WHITLOCK

Copyright © by Alastair Henry and Candas Whitlock
First Edition – November 2014

ISBN
978-0-9939427-0-9 (Paperback)

All right reserved
No part of this book may be reproduced in any form, or by any means, electronic or mechanical, including photocopying, recording, or any information browsing, storage, or retrieval system, without permission in writing from the publisher.

Published by:
Alastair Henry and Candas Whitlock

Printed by:
Double Q Printing & Graphics, London, Ontario

**www.goforitvolunteeringadventures.com**

Cataloguing-in-Publication Data available from Library and Archives Canada.

# Table of Contents

Prologue . . . . . . . . . . . . . . . . . . . . . . . . . . . . . . . . . . . . . . . . . . . . . . . . i

Chapter 1
    Alastair's Journey. . . . . . . . . . . . . . . . . . . . . . . . . . . . . . . . . . . . . 1

Chapter 2
    Candas' Journey. . . . . . . . . . . . . . . . . . . . . . . . . . . . . . . . . . . . . 29

Chapter 3
    Our Journey
        Jamaica. . . . . . . . . . . . . . . . . . . . . . . . . . . . . . . . . . . . . . . . 43
        Guyana . . . . . . . . . . . . . . . . . . . . . . . . . . . . . . . . . . . . . . . . 103
        Tobago . . . . . . . . . . . . . . . . . . . . . . . . . . . . . . . . . . . . . . . . 136
        Guyana (continued). . . . . . . . . . . . . . . . . . . . . . . . . . . . . . 140
        Antigua . . . . . . . . . . . . . . . . . . . . . . . . . . . . . . . . . . . . . . . 146
        Rupununi . . . . . . . . . . . . . . . . . . . . . . . . . . . . . . . . . . . . . . 148
        Guyana (continued). . . . . . . . . . . . . . . . . . . . . . . . . . . . . . 168

International Development Volunteering . . . . . . . . . . . . . . . . . . . . . 171

Epilogue . . . . . . . . . . . . . . . . . . . . . . . . . . . . . . . . . . . . . . . . . . . . . 173

# Prologue

"GO FOR IT" is a story about Alastair and Candas: two Boomers who wanted to help make a difference in the world. We met by chance one day, fell in love, and with a shared passion to help improve the lives of others, set off on a new path of adventure merging our lives in travel and volunteering. "GO FOR IT" describes our personal journeys, before and after we met.

If you're recently retired or soon planning to, and are working on your "bucket list," then we hope our stories will inspire you to "GO FOR IT."

As we think about retirement we contemplate what we'll do, where we'll live and what we'll need, keeping in mind there's a good chance we'll spend more time in retirement than we did in our work life. Will we have enough money to continue our present lifestyle, or to fund the changes to it that we'd like to make? Will we have the funds to realize the dreams on our "Bucket List?" Will we get enough mental stimulation and challenge to remain intellectually vibrant? And what about health benefits, services and family support? – will they be sufficient to manage the aging process, or a serious illness? To sum it up, what will we be: a wealth-schemer, a leisure aficionado, an obsessed penny pincher, or a "GO FOR IT" opportunistic trailblazer?

We Baby Boomers are now the group most targeted for sales and services. We are constantly plagued by social media advertising to buy the right retirement plan, the right insurance policy, the right health benefits package, the right adult life style housing etc., etc. Companies try to convince us they know what's best for us, but you and I both know all they want is our money. They market their products by instilling fear and worrying thoughts that we'll have a dismal, insecure, destitute and guilty filled final stage of life if we don't buy their products.

But what if we strip away all that marketing hype and simply ask ourselves what is important to us, because it is our final stage of life – our last chance to live our dreams.

There's a plethora of options available to us according to our individual interests, situations and resources. We encourage you to think outside the box; to step outside your comfort zone; to be courageous; and to consider what makes you happy. Some of you might already be in a situation that meets all your needs and don't feel you need to make changes. For others though: those looking for a change and adventure, it might mean journeying down an unfamiliar path leading to a destination called, "All that you can be." Your pathway may lead to an address just around the corner from where you live, to a senior's residence across town, or to an orphanage on the opposite side of the world!

Regardless of your situation, we hope you enjoy reading about our journey, and for those of you who are looking to do something different, we hope our story sparks and flames your fire, and inspires you to "GO FOR IT."

We all have a "Start Date" (our birthday), perhaps some "Best Before Dates," and an "End Date," but none of us know when that will be. For us Boomers, the period between now and our "End Date" can be the "Best" times of our lives because we have the circumstances to make it so. We have the wisdom that comes from living on this Earth for sixty odd years, the time to pursue the dreams on our "bucket list," and for most recently retired Boomers, good health, mentally and physically.

To help get you started on thinking outside the box and stepping out of your comfort zone, try this exercise: Go for a drive with no destination in mind. Follow your instincts as to whether you turn right or left at each intersection. Stay in the moment and you'll see, meet and experience something new every time. This game was suggested to us by our friends, Marion and Mike, who bought a home and a car like this. And we're sure many of you, when you reflect on your life's journey, will also remember chance encounters that substantially changed your life.

Everyone's life story is unique and interesting for different reasons. Open yourself to chatting to the person in line at the grocery store, or sitting at the next table at a restaurant, or to the person reading a book in a waiting room. Before you know it, you're chatting away and stories – yours and theirs– are being revealed. And who knows where that might lead? It might change your life's direction, because that's what happened to us.

This book came about through the urging of family and friends to share our stories with others so they might enjoy reading about our journey; be better informed about volunteering overseas; and "if the shoe fits," to be inspired and decide to "GO FOR IT" themselves. Remember this is it. This isn't a dress rehearsal for a more fulfilling life next time around!

# GO FOR IT

CHAPTER 1

# Alastair's Journey

## A Paradigm Shift in Thinking

I first retired at fifty seven when the company I owned went bankrupt. It was earlier than I expected, but as I had an idyllic retirement home in the country – a fifty acre spread with a river running through it, five ponds and eighteen acres of forest, and I could afford to retire, I decided to "GO FOR IT." The retirement honeymoon and the chance to get closer to nature by helping my rural neighbors set beaver and coon traps, bring in hay, feed the chickens and pigs etc., was initially novel and pleasurable, but after a year of leisure, I found myself asking, "Is this it? Is this all there is? Is this all I'm going to do for the rest of my life?" It wasn't enough. There was something missing and I had an urge to find it. This became the focus of my new journey.

To cut a long story short, I returned to the workforce by taking a job in a small First Nations community in a remote, fly-in location on the east arm of Great Slave Lake in Canada's Northwest Territories. Living and working in a strange land for two years with people of a different culture, lifestyle and beliefs caused a paradigm shift in how I looked at the world and what I wanted to do. Along the way, I sold my house and disposed of my furnishings, and in return, gained a deeper and more satisfying sense of who I was, eventually achieving the inner fulfillment I'd been seeking for a long time. I wrote about this search for meaning in my memoir, "Awakening in the Northwest Territories."

When I came out of the North and reflected on what I'd learned, I realized I had a burning desire to use my skills to help others, regardless of whether I got paid or not, and to continue traveling and experiencing different cultures and lands. My three children were independently living their lives and on their own paths, following their instincts, dreams and desires.

## Teaching English as a Second Language

My first plan was to teach English as a second language in a foreign country. This was exciting because I viewed it as a new lease on life, as it were. I took the basic "Teaching English to Speakers of Other Languages" course (TESOL) with a local teaching academy in Edmonton, and chose "Teaching Business English" as my specialty. I applied for positions all over the world, and finally landed one with a University in China, which I subsequently declined after much soul searching and researching the location. Although I had an accounting degree – Certified Management Accountant (C.M.A.) – I didn't have a University diploma to fax over with my application, and that proved to be a major hurdle in securing a teaching position overseas.

However, I wasn't daunted by this development because many English school operators told me replacement teachers were always in high demand. If I was in country, I'd pick up a job quickly, and at a higher rate too, because they wouldn't have to pay my air fare to and from Canada. The other advantage in going to a country under your own steam is you get to physically check out the school, your accommodations, as well as the schools' reputation in the community. That avoided surprises, which could be catastrophic because of the inherent commitment in signing a contract for a specified period of time, sight unseen.

## International Volunteering – Bangladesh

Just as I was thinking about where to go for a teaching position, I became aware of a U.K. based volunteer sending organization called Volunteer Services Overseas (VSO). Out of curiosity, I completed the on-line application and researched what the agency was all about. I don't recall now how I found out about VSO, but it changed my life from that point on.

Within a couple of months, I was on my way to Bangladesh as an Organizational Development Adviser. How profound was that? I smiled when I thought about how much energy, sweat and tears I'd spent earlier in life fretting over trivial matters, thinking they were mega critical, but in the final analysis they weren't. All stuff is small stuff – Right? And here was I, at sixty one, with relatively little research and no second thoughts, going half way around the world for two years by myself to work with some organization I knew little about. Life truly had become an adventure. Not owning a home or having any material possessions to worry about was an added bonus.

In Ottawa, at the VSO pre-departure training workshop, I met two other volunteers, Ramona and Jennifer, who were also going to Bangladesh, but they were destined to work in the hinterland, whereas my placement was in the capital city, Dhaka.

I can't begin to describe how excited and alive I felt with this impending adventure: even getting the physical examination and the vaccinations were thrilling, in a strange sort of way. At the time, I knew little about Bangladesh and even less about Dhaka, but today, with the internet, you can get as much information as you need.

Bangladesh is located in the low lying Ganges Delta in S. Asia, formed by the confluence of three huge rivers: the Padma (known as the Ganges in India), the Jamuna (known as the Brahmaputra in India), and the Meghna rivers, and their respective tributaries. All rivers flow south and empty out into the Bay of Bengal. It's bordered by India on the north, west and north east and by Myanmar (Burma) on the southeast. It's also close to Nepal, Bhutan and China, and for wannabe wanderlust such as I, that was exciting.

Bangladesh is a developing country with a turbulent history of political turmoil, corruption, assassinations, flooding, poverty, famines and military coups, but that didn't faze me. I wanted to go there to help whomever and wherever I could, and if anything were to happen, I had complete confidence VSO would look after me, and they did. But if for whatever reason, I did perish over there, so be it. We all die somewhere from something, do we not? I realized I wasn't afraid of death any more, as I had been earlier in life, and that was a calming realization.

A hundred and fifty million people live in Bangladesh and seventeen million live in Dhaka, along with four hundred thousand cycle rickshaws, thousands of mosques, and billions of mice and cockroaches.

I vividly recall how I felt when the VSO vehicle drove us from the airport to the orientation guest house: the experience was so overwhelming and alarming it's forever etched in my memory: I can see masses of humanity jostling for space on every street and hear the harsh cacophony of car horns, whistles and shouts; and smell the revolting air leaking into the van as we sped along. Yet I also recall feeling excited as the energy and vibrancy of the place seeped into my psyche. I was as surprised and pleased as if I'd scratched a winning lottery ticket number and won a big prize.

## Dhaka – First Impressions

VSO's policy was for volunteers to live at the same economic level as their work colleagues so as to give them an authentic experience for what daily life was like. As well, it aided in the acceptance of volunteers by staff, by leveling the economic playing field as it were. You lived close to where you worked, in similar type housing as your co-workers and with the same amenities as enjoyed, or in the case of volunteers, "endured" by them. This was in sharp contrast to staff hired by large International Development Corporations, who offered attractive salary and benefit packages to compete globally to entice workers. They typically provided fully equipped accommodations in the safest areas of the city, rides to and from the workplace, servants, cleaners and cooks, and in some cases, a car and driver.

My home was a third floor flat in Lalmatia. There was no hot water or air conditioning, but there were plenty of mice and cockroaches scurrying around the cement floored kitchen. My room-mate, Edison, a Ugandan from Kampala, worked with five other Ugandans on a VSO HIV/Aids prevention program in the country.

When Edison's placement ended and he returned to Uganda, I moved over to share Harrison's flat – he was Ugandan too – and I wasn't surprised to see he had as many mice and cockroaches as us. It's just the way it was because you couldn't permanently get rid them. You just had to spray the rooms for cockroaches and set traps for the mice. What worked best was glue and peanut butter. You oozed out a circle of glue on a piece of cardboard and put a dollop of peanut butter in the middle.

My work place was in Mohammadpur: a thirty minute daily commute by rickshaw.

When a cohort of volunteers first arrives in country, they're housed together for five days of intense orientation workshops and language training. There were fourteen in our batch: three Canadians, six Filipinos and three Ugandans. Fourteen two-hour language lessons to teach us Bangla were scheduled, but because of political unrest – protest in the streets – we only attended ten. That was enough for me because I was based in Dhaka and the staff wanted to practice their English to strengthen their linguistic ability and resumes, but that wasn't the case for Ramona and Jennifer. They had a much tougher time with the language because they lived in the country-side, where few people spoke English.

I have to admit I initially felt apprehensive about learning a new language, given my advancing age and past memories of doing so poorly in Latin and French at school, but I applied myself and thoroughly enjoyed the experience. I didn't become bilingual of course, but I did retain enough words and phrases to get by. My work colleagues appreciated my efforts to converse with them in their own tongue and they giggled at my pronunciation.

Bangla is a simpler language than English: at least that was my impression. It seemed to have fewer words in its vocabulary. Everyone greeted me with, "Kemon achen," pronounced "Kay mon ah shen," meaning "How are you? How's it going?" And the standard response, if everything was okay was, "Ami bhalo achi," pronounced "Ah me bha lo ah chee." I can imagine how difficult it must be for non-English speaking people to understand the hundreds of ways we have of greeting each other and of responding. "Bhalo" was used to describe everything that was good, whereas in the English language, dozens of words express that: great, fantastic, super, brilliant, wonderful, splendid, fantastic etc.

What I appreciated most about VSO's approach to international development was their focus on capacity building, through facilitating rather than doing. As a volunteer, you try to change lives by helping to teach people how to fish for themselves, and not by catching fish for them. It was all about sustainability. Many times I could've just done the work and moved on, but I arranged a work shop instead to get my colleagues involved. Their sense of ownership of the work was what it was all about.

When I was sixteen and leaving school, I wanted to be a teacher, but never got the chance. Now here I was, forty five years later, teaching, and thriving on it. I didn't see that one coming.

In workshops, when we were asked to demonstrate various aspects of life in our own country, the Ugandans always had demonstrative and interesting scenarios to act out. And invariably, their show was always about the subservient relationship women had to men. They took great pains to explain the relationship was all about respect and recognition of each other's role in the family, contrary to how we in the western world negatively view subservient relationships of any kind.

Volunteers, by their nature, are kind, caring and understanding, and these traits are extended to fellow volunteers in addition to the people they work with in their placements. When they leave their placements, they typically give their possessions, such as saris and boots, as well as household items like pots and pans, water filters etc., and left over groceries to people they know could use them.

## SAP-Bangladesh

My NGO was SAP-Bangladesh (SAP): an organization that worked in the rural development field through capacity building of individuals, communities and local level small NGOs, in the remote coastal and riverine districts of Patuakhali, Barguna, Bagerhat, Sirajgonj and Gaibanda. Although I was based in Dhaka to help the staff develop a five year strategic plan, I fully expected to visit each district office at some point over the next two years and that was an exciting prospect.

## Micro Credit

The abjectly poor don't have bank accounts or credit cards. They don't have collateral, steady employment or a verifiable credit history and therefore cannot meet even the most minimal requirements for a traditional loan. Professor Yunus pondered this situation, and in 1976, in the village of Jobra, Bangladesh, he developed a system called Micro Credit. Over the last thirty years, micro credit has enabled millions of people all over the world to break out of the poverty cycle and become financially independent. In recognition of the beneficial impact micro credit has had in lifting people out of poverty, Professor Yunus was awarded the Nobel Peace Prize in 2006.

Micro credit is based on a set of principles, different from general lending criteria. It builds the capacity and trust of micro-entrepreneurs, encourages employment generation, teaches financial management and helps its members through difficult times. Repayment rates are high because of peer pressure and support. Borrowers are responsible for each other's success and this seems to ensure that everyone repays their loan. In Bangladesh, over $4 billion was loaned out as micro credit loans between 1996 and 2006; in India, in 2006, $1.3 billion was loaned out to 17.5 million people; and in Latin America and the Caribbean, $8.6 billion was loaned to 8 million people in 2007.

It works like this: Villagers form small micro credit groups, appoint a treasurer, get a pass book and meet weekly to conduct business.

Each member saves a small amount of money and the group democratically votes on which member gets the pooled amount as a loan. For example, let's say there were twenty villagers in the group and they agreed to save $.25 per week, or $5.00 in total. Member A might want to buy some seeds, member B a chicken, and member C a fish knife. The group has $5.00 to loan out every week, and after hearing each member make a case for why they want to borrow the money, the group decides who gets the loan. The Treasurer records each transaction in member's pass books, and calculates interest earned on savings and interest charged on loans.

SAP had an extensive and successful micro credit program for the villages they served. I'll tell you the story about one couple in Patuakhali that I became aware of to give you an idea of how powerful micro credit can be in changing poor people's lives.

Amena's marriage at the age of thirteen to a much older man was arranged by her father. Her parents were separated and her mother was living "on the street." By the time Amena was sixteen, she had borne two children. At eighteen, she left her children and abusive husband, and went to live "on the street" with her mother. Molla was an amputee – he'd lost one leg – and survived "on the street" by earning tips from cleaning the pay public washrooms. Molla took a liking to Amena and watched out for her and her mother.

One day, Amena attended a SAP meeting on micro-credit in the village and came away enthused and hopeful that it might be the way to get her aging mother off the street, which was what Amena most wanted to do. She discussed the concept with Molla, and they decided to join the group and ask for a loan to buy coconuts and a machete. Molla would sell the coconuts to passers-by, while waiting outside the washroom for tips. They got their loan and repaid it on time. They then got a larger loan, Molla bought a bigger supply of coconuts, and Amena opened up her own stall. They saved hard, and when the pay public washroom lease came up for renewal they applied and were successful. It was a happy ending because they eventually earned enough money to rent two rooms in the village, one for them and one for Amena's mother.

## Black Tea

SAP staff was exceedingly welcoming and I settled in quickly. I had a desk, computer, and access to whatever supplies I needed. Many volunteers didn't have it that good. Every mid-morning and afternoon, the "tea-boy," a young man in his middle twenties, brought me a cup of boiling hot tea with a dollop of carnation milk and two heaping teaspoons of sugar, despite my informing him every day I took my tea "black." He just couldn't bring himself to pour a cup of 'black tea" because that wasn't the way you took tea in Bangladesh! And with the tea came a china plate holding three or four sweet biscuits. How civilized was that, and how privileged was I?

In no time at all I'd adapted to my new way of life: waking up at 5:00 a.m. with the calls to prayer from the many mosques surrounding my flat; showering in refreshingly cool water; eating loads of rice and chicken; hailing rickshaws to take me to and from work; and even learning how to squat to use the hole in the floor as a toilet! I was pleased with my contribution at work and my relationships with colleagues, and I enjoyed the many social chit chats I had with staff. They couldn't understand what I was doing there at my age, when I had three children back in Canada with whom I could live with if I didn't have a wife!

All the staff were Muslim and a few went into the prayer room at work to say their prayers at the appointed times during the day. Their meekness and sincerity impressed me. They quizzed me on my religion and were surprised to learn I wasn't a practicing Christian – they thought all Canadians were – but more importantly, they were astonished to learn I didn't belong to any one religion. Needless to say, we had many interesting spiritual and philosophical discussions about that.

Another great dividend to international volunteering is the convivial friendships you develop with volunteers from other countries. Because you share personal space for prolonged periods of time, from one to two years, you usually end up bonding with your flat mate on a deep level, which could be further enhanced if there were situations that brought you closer, which is what happened to Harrison and me when I got dengue fever.

## Dengue Fever

Harrison looked after me when I was "out of it" for four days and truly didn't care whether I lived or died. The fever puts you in a delirious state, causing you to vomit and run to the bathroom every half hour. It seems to stop your rational mind from thinking, leaving you zombie like, where you just exist and aren't aware of anything. Harrison came into my room every few hours for three or four days with a tall glass of water, sugar and salt. He stayed and watched to ensure I drank every drop. I shudder to think what would have happened had I lived alone and no one was aware of my condition. I'd not be here writing about my experience, that's for sure.

## Staphylococcus Aureus

I recovered quickly once the fever passed, but it left me with a rash on my lower back that got infected, apparently from my bed sheets. I subsequently learned I should have hung my bedding on the balcony rail every day, exposing it to sunlight to kill the microbes. Initially, I thought the rash was just an itchy irritation caused by sweating from the jiggling of my back-pack on my lower back whilst traveling on the rickshaw every day. When my lower back began to harden and swell, I went to Elizabeth House, the medical center appointed by VSO, to find out what was going on. The doctor didn't know specifically what I had, and so she gave me penicillin tablets to see if that would do the trick and told me to return in a week.

The hardening increased daily resulting in my having to sleep on my back or front, but not on my side, and by about the third week, the swelling began to manifest in horrible and painful carbuncles and boils. That was a good development because the doctor could now take a swab and diagnose what type of infection I had. It turned out to be the staphylococcus aureus, a nasty bug that could become a flesh eating disease if left untreated. It was resistant to the penicillin I was taking so the doctor changed my medication, gave me a week's supply of bandages, dressings and tape, plastic gloves, antiseptic solution and a medicated cream for bathing the wounds, and set up a schedule for me to visit the Clinic every Thursday morning.

My challenge was to find people to do the dirty work of changing the dressing twice a day for the first month, and then once a day for another three months, which was how long it took to fully recover. Between fellow volunteers and work colleagues, a total of fourteen individuals worked on my back. A few declined, saying it was too gross a task for them, but surprisingly, most people volunteered their services when they realized the importance of changing the dressing. The doctor was concerned about my immune system because the strong antibiotics I was ingesting also killed the good and necessary bacteria in the body, and if I was to contract another dose of Dengue, it would probably be the hemorrhagic type and that would be bad news.

## The Bagha Club

Lalmatia, where our flat was located, was a suburban residential area where most of my work colleagues lived. Forty minutes away by a compressed natural gas three wheeled vehicle (CNG) was Gulshan, the best area of Dhaka: the cleanest, safest and most modern district. This was where the foreign embassies were located and wealthy families lived. There was a gated facility there called the British Aid and Guest House Association Club (BAGHA), a remnant from past colonial days I think judging by its style and decor. It had a restaurant that sold western type food, a swimming pool, tennis court, and a bar with a big screen TV, which showed English Premier League soccer games on the weekend. It was a refuge for foreigners working in Dhaka and so served an important role as a social meeting place for ex pats.

As Bangladesh was a Muslim country, alcohol wasn't publicly available in the country, not even in restaurants, so the availability of alcohol at the BAGHA was greatly appreciated by all ex-pats. VSO negotiated a special membership fee with them for its volunteers. The cost was quite affordable, so most volunteers who were based in Dhaka, as well as other ex pats working in Bangladesh became members. The opportunity to disappear once a week into the BAGHA for relaxation and a taste of "over home" was wonderful, and no doubt responsible for many volunteers staying the course, who otherwise might have terminated their placements early.

Being based in Dhaka, as opposed to a placement in a rural or coastal village, was a tough assignment in many ways because of the sheer number of people living there and the corresponding intensity of daily life – the traffic, noise, pollution, that sort of thing. By the end of the week I was usually tired and a little irritable.

Every Friday (Muslim Holy day) I went to the BAGHA after work by CNG and sometimes on a Saturday too, if I felt in need of a swim, a regular meal, or a beer. I made sure to hail only a white taxi for my trip home, because VSO said they were the most reliable. They said we should never, under any circumstances, ever get into a black taxi because they were notorious for muggings and for conning passengers into paying more than the "going rate."

Another enjoyable benefit to BAGHA membership was in meeting people working for different companies. This resulted in most interesting conversations and new friendships. One volunteer met her future husband there: a young Australian guy who'd been hired by the Bangladeshi cricket team as their physiotherapist.

I met Dave and Carol. Dave was from my home town in England, was about the same age as me, and had a similar childhood growing up in Lancashire. He was an educational consultant, one of many from around the world, who were working on a big five year project for the Bangladeshi government. He was highly paid, had a comfortable flat in Gulshan with hot water, air conditioning, a maid to cook and clean, and a car and driver. They also had an extra bedroom, which in time became "Al's room," but not before a scary episode brought about the idea of me staying over on a Friday night and being driven home on Saturday morning by their driver.

# The Robbery

About a month after I joined the BAGHA, I hired a white taxi as suggested by VSO to take me home. It was about ten o'clock at night when I negotiated the right price with the driver at one hundred taka (about $1.50 Cdn). I settled down in the back seat for a relaxed ride home, for it being a Friday night, it was the quietest night of the week. About ten minutes into the ride, the driver pulled over onto the soft shoulder, intimating he had a problem with a headlight. No problem. I was prepared to wait patiently for him to fix whatever was wrong, but then my world turned upside down.

Three scary characters mysteriously emerged from the dark bushes and climbed into the car: one on either side of me in the back seat, and one in the passenger seat at the front. The driver sped off while my assailants frisked me. The guy at the front examined my day pack containing a phone, MP3 player, camera, sun glasses, swimming trunks and a towel. The guy on my left put a choke-hold on me and forced me back into the seat. I blurted out that there was no need to do that for there were four of them and only one of me! He obviously understood some English because he immediately released his grip and proceeded to pat me down all over, including under my belt and around my thigh and calf. The guy on my right undid my shoe laces, pulled off my shoes and inspected my socks.

The driver made a U turn, came back and stopped at the top of an overpass. He asked his accomplices what they'd found. The guy in the front, who went through the back-pack, said there wasn't anything. The driver screamed at me, "Cell phone? Cell phone Na?" I shrugged my shoulders intuiting that the best course of action was to say and do nothing. The guy on my left, who relieved me of a thousand taka – one five hundred taka note and five one hundreds – showed only the five hundred note (about $7.00 Cdn) to the rest of the gang. The driver couldn't believe this was all I had. He continued to scream at me, "Cell phone? Cell phone na?" The guy on my left opened the door, got out, and came around to the right side. Both attackers on the back seat then pushed me to get out of the vehicle.

I didn't need to understand Bangla to know what was going on. They were dumping me on the highway and going to drive off. I didn't budge. I complained I needed a hundred taka for a taxi home. They understood my every word, but didn't offer me any money.

They kept pushing me, but then a huge transport truck came up behind us and stopped. I didn't turn my head to see, but I knew what it was by the heavy rumble of its engine and its bright lights shining directly into our vehicle. The guys in the car continued to argue intensely and then the driver floored the accelerator and took off, careening down the overpass. He did another U turn and returned to the spot where his accomplices first got in. The shoving and shouting continued, and I yelled back at them I needed money to get home. The guy in the front seat handed me my back-pack, and the one who'd taken the money gave me a hundred taka note. I got out. They pulled the door shut and sped off at high speed.

I was dazed and shaking as I waited in the dark on the lonely highway for some good soul to come along and help. Imagine my chagrin when the first vehicle to appear and stop was the dreaded black taxi: the one VSO said never to take under any circumstances! Could my situation get any worse? The driver could obviously see that something was seriously amiss, and when I told him what had happened, he agreed to take me home for a hundred taka.

I went to bed quietly and quickly fell asleep. The next morning, Harrison was stunned I hadn't woken him up when I got home to tell him about my ordeal. He said I must have been traumatized and not thinking straight and he was right, up to a point. I was in shock naturally, but I also understood that these things happen, even in my own country, and I was okay with that. I wasn't hurt, and at no time did I feel my life was in real danger. I'd been robbed of some stuff and about seven dollars in Canadian funds, but that was it. I reasoned that the guys needed the money to feed their families, or to pay some hospital bills, or school fees. I even thought they might have needed the money to buy their next fix, and I was okay with that too. Who was I to judge when these people had so little and I so much!

Dave and Carol were upset and angry with Bangladesh when they heard what happened. They insisted I stay over every Friday night and have the second bedroom in their condo, and then be driven home in their car by Dave's driver the following morning. Well, I gotta tell you, it was like dying and going to heaven. Dave and Carol were so warm and generous with their hospitality that I soon forgot about the rigors of living so spartanly the other five days of the week.

## A State of Emergency

I didn't buy a replacement cell phone or MP3 player. Why bother when I could be relieved of such items simply by getting into a white taxi. Besides, VSO said cell phones weren't an essential item, and that was why they didn't provide them, but they also said having a phone for emergencies was a good idea. I questioned this policy soon after arriving in country because many Ugandans and Filipinos didn't have any extra money when they arrived. They used their first month's food allowance to buy a phone. I didn't think this was fair and I raised the issue with VSO, but they insisted a phone wasn't an "essential" item.

A few months later, when a potential state of emergency arose due to elections in the country, VSO arranged an emergency evacuation drill, just in case. They sent all volunteers a text message and an e-mail to immediately proceed with the evacuation plan – Stage 1, which was to proceed to the appointed "safe house," and to wait there for further instructions. As it turned out, I was on my way to the VSO office at the time to get my e-mails. Before I had a chance to login, the Country Director appeared, and all hot and bothered, asked what I was doing there.

"Getting my e-mails," I naively responded.
"Why aren't you at a Safe House?"
I had no idea what she was talking about.
"Didn't you get my message?"
"No."
"Don't you check your phone messages and e-mails?"

"I don't have a phone, and I don't have internet at home – that's why I'm here.

"What's up? What's happened?"

"Why don't you have a cell phone – you need a cell phone."

And then a look came over her face as she recalled our previous discussions about whether a phone was essential or not. I felt bad because her policy had back-fired and put her in a sensitive and embarrassing situation. The feedback from the evacuation drill assessment was that all volunteers should have cell phones, but it didn't go any further than that. I don't know if cell phones are now given to volunteers upon arriving in country, but I hope so.

## "There, but for the grace of God, or Allah, or whoever, go I."

As I watched "older" people – many much younger than me – crippled with some form of physical disability, beg outside rich people's homes, the same thought always came to mind: "There, but for the grace of God, or Allah, or whoever, go I." One particular situation I saw quite often that always resonated with me was a small, wooden cart with wooden wheels, holding four older men with amputated limbs, being pulled by a boy aged no more than nine or ten. He reminded me of my grandsons. They stopped outside large houses and the men chanted a prayer to Allah. Invariably, the gate to the complex would open, and some good soul from the kitchen would come out with a food tray of vegetables – cut off ends of carrots, outside leaves of cabbage and lettuce – that sort of thing.

I recall feeling overwhelmed that I'd been born into circumstances so fortunate. I had no say in who my parents were, or in what country I was born, and nor did these beggars. It was just the luck of the draw of life that they were there and I was here. This experience and realization increased my appreciation for Canada and of how fortunate all of us are who live in the West. That motivated me to want to help more.

## Amena and Molla – Incredible Indian Adventures

In September of my first year, I took a ten day vacation in Incredible India, starting with an overnight bus ride from Dhaka to Calcutta, where I met up with Ramona and Jennifer. Jennifer went on to an ashram while Ramona and I traveled, by overnight trains mostly, to the Taj Mahal in Agra: the white marbled mausoleum built by Shah Jahan, a Mogul Emperor in central India, for his wife, Mumtaz Mahal, who died in childbirth with their fourteenth child. We also visited nearby Fatehpur Sikri and the Red Fort: a walled city built from red sand stone by Emperor Jahangir, comprising royal palaces, harems, courts, a mosque, private quarters and many other buildings. It truly was incredible, and never in my wildest dreams did I ever think that one day I'd see those sights with my own eyes. It was unforgettable.

So too was Varanasi: India's spiritual and cultural capital for thousands of years. We traveled to this most holy site on the banks of the Ganges River by overnight train from Agra. Hundreds of ghats (steps down to the river banks), shrines, temples and palaces, built tier on tier, line the river frontage for many miles, and at night, open cremations take place in front of some of the Hindu temples. Sick, old people from far and wide go there to die and be cremated.

We hired a boatman to take us out on the river in the early morning to experience a sunrise over the Ganges, where we were most fortunate to see the seldom seen river dolphin. Our man said we must be "most blessed" for the dolphin to show himself to us. He took us again that night to witness from the river the incredible audio/visual spectacle happening on the ghats. Thousands of pilgrims chanting, praying and milling around a dozen funeral pyres blazing away on the steps; thousands of floating candles being released for their journey to the sea, bobbing in the water between bathers and boaters; as well as tourists from all over the world mingling with the crowd on the shore and in the water. We idled in front of one ghat to watch a wrapped body be placed on top of fiery embers, and the fire being stoked with new wood to blaze away and reduce the corpse to ashes.

The following day, we hired a rickshaw to take us to Sarnath, thirteen kilometers away. Sarnath was the deer park where Guatama Buddha, after his enlightenment, first taught the Dharma to his followers. It was a mystical place with a diverse array of temples, built over the years by Buddhists from many lands. It was of special significance to me because lately I'd been reading much about the Buddha, was in the habit of meditating, and found the Dharma teachings to be useful as a framework for living a peaceful life.

Six months later, I went on another holiday to India with three VSO colleagues from the UK to see the "living bridges" in Cherrapunjee: the wettest place on earth. To get there, we flew from Dhaka to Sylhet in northern Bangladesh, walked across the border into the state of Meghalaya in India, and got into the waiting vehicle of the owner of the Cherrapunjee Holiday Resort.

The "living bridges" were the tree roots of an Indian rubber tree that the Khasi (the local indigenous people) trained to span streams and rivers, and to take root on the other side. They intertwined the roots as they grew, thus forming a secure bridge that can hold fifty or more people at a time. The bridges varied in length from fifty to over one hundred feet, and some had two protective railing spans, also created from trained tree roots. The Khasi, who live in little villages on the banks of the densely covered hills, use the "living bridges" daily to commute from one village to another.

To fully describe my adventures in India would be a book unto itself. I mention them here to illustrate what affordable, exotic vacation perks are available once you're working in a different country.

I returned to Canada three months early in my placement as a precautionary health measure because the doctor was concerned about my immune level. She wanted me to go home and get strong. I initially thought I'd hang in there for the full term, and I told everyone that was what I'd do, but for some strange reason, I awoke one morning with a strong conviction I needed to go home. It wasn't a rational, intellectual thought: more of an intuitive, emotional one, which was difficult to explain to people when they asked why I'd changed my mind.

*Al at the Taj Mahal*

*Bathers in the Ganges at Varanassi*

*Cherrapunjee (wettest place on Earth)*

*Living root bridges*

## Grandchild Care Giving

Upon my return from Bangladesh, I lived with my daughter, Nicole and her three children. My two other children, sons Darrin and Dean and their families, lived in different towns about an hour away. When I arrived, a few days before Christmas, I learned of my daughter's plans to return to school early in the New Year. There was a problem though in that she didn't have a caregiver for Beckett, her one year old son. She'd registered him with a number of Day Care Centers, but they had no immediate openings for him. As her course started on January the 4th, I agreed to be the little guy's caregiver until a spot opened up. I canceled my March departure plans to go to a VSO placement in Rwanda, and expected that by June, I'd be off again to some far flung destination in the world, but it didn't work out that way.

My daughter successfully completed her course, but she couldn't find a suitable job in town. It was the same old problem – she needed past work experience for the jobs she applied for. To get her foot in the door, she took a job in her field in another city two hours away and stayed at a hotel for two nights on every shift. As I'd no firm plans, nor any fixed abode, I willingly agreed to continue the caregiving indefinitely, as well as look after the house, her two other children, ages ten and twelve, the dog and a cat.

It was intense at first because for many years I'd lived quietly, mostly on my own. But once I got into a routine, the rest just fell into place. In fact, the children's energy, exuberance and laughter rejuvenated me. Not that I was looking for a purpose in life because I wasn't. I felt complete and at peace with myself and what I was doing, but this new opportunity presented the chance for me to play a part in a young child's development, as well, get closer to my other grandchildren.

Three of my grandparents passed away before I was born, and I only saw my grandmother, on my mother's side, twice. As a child, I recalled feeling I'd lost out on something special because my friends seemed to all have grandparents and had such wonderful times with them. The truth was in those days grandparents did play a big part in the rearing of their children's families, mostly out of necessity. And besides, it was typical for family members in Bolton, Lancashire to live close to their children, even on the same street or next door.

As the weeks and months passed and the little guy began to walk, speak and accomplish certain tasks, I looked on in awe: amazed at how quickly human development occurs in the young, and was surprised with the deep love growing in my heart towards him. I'd forgotten how wondrous it was to have little ones around. It had been forty odd years since my children were this age, and at that time, many distractions had occupied my mind. As a grandfather, it was a totally different experience. I shared in his joy of being; smiled at his innocence and naivety; marveled at his deep sleep; treasured his blind trust in me; and relished his firm, sincere hugs and sloppy kisses. It was like watching a bud blossom into a flower.

## *I Took a Piece of Plastic Clay*

I took a piece of plastic clay
And idly fashioned it one day
And as my fingers pressed it still
It moved and yielded to my will

I came again when days were past
The bit of clay was hard at last
The form I gave it, still it bore
And I could change that form no more

I took a piece of living clay
And gently fashioned it day by day
And molded it with power and art
A young child's soft and yielding heart

I came again when years were gone
It was a man I looked upon
He still that early impress bore
And I could change his heart no more.

---

Was it my daughter's situation in October that beckoned me on an intuitive or psychic level, to return to Canada? One has to wonder. It truly was serendipitous to have events unfold in such a way to allow me the opportunity to learn more about the miracle of life that had been all around me for the last sixty years, but which I'd taken for granted.

It was during this period that I first became aware that my long term memory was improving. It was the weirdest situation. I began to remember in detail most of my whole life, as if I'd been given the key to the library in my mind. I began recalling faces and names of kids I went to Primary and Grammar School with; names and faces of colleagues I worked with, thirty and forty years ago; as well as general information I thought I'd forgotten about years ago. Just when I expected my memory to deteriorate from the aging process, it was getting better. How strange and wonderful was that? The only explanation that I can muster is that the result of having mental clarity from a peaceful and uncluttered mind.

## Backpacking with Grandchildren

My daughter eventually landed a job in her line of work in the city where she lived, and a Day Care spot opened up for my grandson. I continued to live with them and to help out wherever I could because she worked ten and twelve hour shifts, and it was a busy household. One day, Kiefer, my twelve year old grandson, came home from choir practice and announced the choir was going to Prague, Czechoslovakia to participate in a musical festival, and then on to other choral events in Austria and Germany.

I considered a parallel tour with some parents and was on the verge of signing up when the Universe sent me another idea: why not use the opportunity to back-pack Europe with him and his ten year old sister, Madigan, when his tour ended? And that's what we did.

My deal with them was I'd pay for their rail tickets, food and lodgings, and they'd pay their airfare, the cost of special events they wanted to attend, as well as whatever gifts they wished to buy.

The enjoyment started immediately as we considered the possibilities of where to go and what to see. We settled on going for a month: three weeks in Germany, France and Italy, and a week on some island in Greece. We didn't want a fixed itinerary with hotel and travel commitments: we just wanted to come and go as we pleased. It was to be a grand tour on a tight budget to see the highlights of Europe.

Because we just back-packed, stayed in the cheapest rooms, washed our clothes by hand or at a local Laundromat, and ate modestly at local cafes, the total trip cost wasn't excessive. I don't recall now specifically how much it cost because it wasn't important. The experience was priceless. There was no other way I could have bought the blissful sense of joy we shared in our four weeks together. What would I have done with the money instead? Bought a new car? Updated my wardrobe? Joined a golf club? Left it in the bank to accumulate? The children will forever remember in their hearts and minds the special times we spent together, and the education they received about the wider world will no doubt impact them in ways only time will tell. They realized their dreams to see an opera in Verona; boat through the canals of Venice; swim in the Med; and gallop across the beach on horseback in Greece. And I have the lasting joy of knowing I helped make their dreams come true.

## International Volunteering in Nigeria

In 2008, a Canadian volunteer sending agency called Cuso merged with the Canadian arm of VSO to become Cuso/VSO. In 2011, the organization unmerged and changed its name to Cuso International (Cuso). My next assignment was with them. It was short term –just one month in Ibidan and Abuja in Nigeria. My task was to help an NGO ensure they had the necessary systems and procedures in place to satisfy the requirements of a major Canadian donor. Even though it was only a fleeting visit, it gave me a small taste of Africa, and ignited in me a desire to return one day.

## Volunteering in Canada

While I was waiting for another appropriate international assignment, I volunteered my services locally. The Northwest London Resources Centre (NWLRC) was looking for additional Board members, and the only requirement was to attend monthly Board meetings. As I was going overseas at some point soon and didn't want a permanent full time position, this situation interested me. I visited the Center to inquire about volunteering and was interviewed by a Ms. Candas Whitlock, the Executive Director. By meeting's end, she offered me a position on the Board.

I joined the Board; worked on their financial budget and five year plan; served first as their Treasurer; and then as Chairman when the incumbent resigned. The work was satisfying and I enjoyed the camaraderie of working with the other Board members and staff.

## International Volunteering – Flores, Indonesia

I resigned from the Board at the end of March to get ready to go to Flores, Indonesia in early June with Cuso for two years. There was an NGO there whose focus was micro credit. They'd set up micro credit loan programs in many villages; had been hugely successful; and were now looking to expand their services across the island. I liked the placement's content a lot and began to do my due diligence. Flores was east of the islands of Sumbawa and Komodo, west of Lembata, northwest of Timor and south of Sulawesi – how exotic was that?

I'd never been to S.E. Asia, but it was on my "bucket list," and this was my big chance. I was eager to go, and even mapped out where I'd spend my vacation. A week's beach holiday in Bali and two weeks back-packing Vietnam, Laos and Cambodia in the first year, and in the second year, I planned to go to Darwin, Australia to check out the Great Barrier Reef and stop on the way back in Flores to see the komodo dragons. It was all very doable.

Two weeks before my planned departure, Cuso called and said there was a hiccup. I don't know specifically what the problem was, but it had to do with the Indonesian government and Visas. They said they needed a couple of months to sort things out, and asked if I was okay with that. I was disappointed of course on one level, but I now accepted whatever was, and so I was fine with it. If it was to be, it would have happened, and if it wasn't, it wouldn't. I agreed to stay linked to the placement at least until September, and used the opportunity to take six grandchildren on a two week camping adventure. I took them up the Bruce Peninsula and by ferry over to Manitoulin Island in the first week, and then to the Dunes at the Pinery Provincial Park on Lake Huron to tent camp for the second week. I was mindful I'd soon be off again for two years and so spent as much time as I could with the grandchildren doing fun things. I also visited the UK for two weeks to visit my brother and sister, and a friend from high school, who I hadn't seen for close to fifty years. We'd hooked up through a web site about a year earlier.

## A Change in Plans and a Soul Mate

In the meantime, I'd become a good friend of Candas. She called one day and invited me to be her guest at a wedding. I couldn't remember the last time I had a "date" – it must have been at least five years. We had fun and began to spend more time together. We discovered we shared the same life philosophies and interest in travel and international development. She too was eager to go overseas with an organization, such as Cuso, to work with the under privileged, and felt confident that with her background in social work, she'd find a suitable position. She applied to Cuso, resigned from her job and put her house up for sale. They accepted her and began offering various placements as they came up. If she got a placement in Asia, we planned to vacation together because back-packing through Vietnam, Laos and Cambodia was also on her "bucket list."

When Cuso advised in September the Indonesian government was still playing hard ball on the visa issue, and it didn't look promising, I asked them to unlink me from the Flores placement and find me something else. I was most eager to go overseas again and to resume my volunteering lifestyle.

Just before Christmas, Cuso offered me an interesting placement in Kingston, Jamaica. Jamaica wasn't as exotic as Flores, but it had many advantages. It was only a four hour flight from Toronto in case of an emergency, there was no need for language lessons, the beaches and Blue Mountains were accessible for weekend "get-a-ways." And most important of all, I could embark on a new chapter of life with Candas for she too had been offered a position in Jamaica with the same start and stop dates. How lucky was that? We were over the moon with joy. It was like being twenty years of age again and planning to move in together! But this time the challenge was how to tell our children, not our parents.

Sometimes in life things happen that seem inconsequential at the time, but turn out to be monumental and dramatically change one's life going forward. This was such a time. Many times I wondered how my journey would have evolved had I not come across the VSO information when I did, or volunteered at the NWLRC.

# GO FOR IT

CHAPTER 2

# Candas' Journey

When I look back over my life I'm filled with gratitude. I've lived in privilege, and with each new day, I thank the Universe for providing the opportunities and experiences that were offered to me along my life's path. This is the way of life – we become all that we can be, in the fullness of time, because of lessons learned throughout our journey.

## From the Beginning

My mother was the most influential person in the development of my appreciation for nature and in my desire to explore the human ways of being. Her open mind, acceptance, generous care, concern and love of people instilled in me a natural curiosity about life. Sylvia Hazell was known as "Sylvie" to friends, neighbors and local shopkeepers, and as "Mama-mia" to her grandkids and neighborhood children.

When she was forty five, she gave birth to her seventh child and named me Candas Faith. When I was old enough to understand, she told me she named me Faith because she wanted me to always remember to have faith in myself. At some point, she gave me a small clear glass marble with a mustard seed in the middle, and these wise words to live by: "If you have the faith of a grain of mustard seed, nothing will be impossible to you."

Soon after my birth, my father abandoned us, leaving my mother a single parent of seven children. My mother said I was born "very sickly" and she feared I might not survive. I don't know what specific ailments I had, other than asthma, but when I started school, I went to a "Health School" for children with special medical needs. In 1947 there was no social safety net to assist families living in poverty, or single parent families. The money coming into our house was meagre, but my mother was industrious – she worked as a house cleaner and sold home-made bread on the street to make ends meet. I don't recall her ever complaining. Moreover, she always had a song on her lips, a smile on her face, and a unique laugh that everyone recognized.

She was only about five foot two, full breasted with a round body that smelled of Jerkins lotion at night and flour during the day, and the only clothes I recall her wearing was a flowered, cotton house dress and an apron

As the youngest child, I was attached to my mother's side wherever she went, and up until her death, we shared the same bed. I loved to trace the dimples on her elbows as I lay curled into her back when we slept, and brush her long red hair, and braid and twist it around the top of her head securing it in place with bobby-pins. I watched her bake bread: her arms jiggling as her small, chubby hands kneaded the bread dough with a rhyme and skill that only comes with years of practice. Her bread had an intoxicating aroma that made your mouth water.

As she got older and troubled by arthritis in her arms and legs, I bathed her feet and rubbed cream into her joints. And even though her teeth were decayed, they didn't stop her from smiling, or speaking to people she met on the street. Her laughter was contagious. Everyone knew when Sylvie was in their midst.

Whenever I think about her, the same image always comes to mind: she's wearing a house dress; her arms are folded across her abundant chest; and she's leaning against the heating wall vent in the living room, warming herself from the coal furnace embers in the basement. Her presence at that very spot eventually wore away the wall paper and left a bare mark on the wall.

My upbringing was very different than my three oldest siblings, Roxie, Jack and Shirley Ann. Although they lived through the depression, they had a full family life experience with a mother and a father, grandparents, as well as aunts, uncles and cousins. Five years after Shirley Ann was born, my mother had Leah, and eighteen months later, Judy. Their childhood was somewhat similar to my three oldest siblings, because dad was still in the picture. My brother Ray was born eight years after Judy, and I came into the world two years later. There was a twenty-one year age difference between myself and my oldest sibling. Ray and I never met our grandparents and had little or no interactions with the extended family of uncles or aunts. Our childhood revolved around our mother and our siblings.

My oldest sister, Roxie, was already married when I was born and I became an aunt when I was only ten months old. Most of my siblings went to work after grade school or when they turned sixteen, to financially help out the family. They married early and started a family, which in the 1950s usually happened around the age of eighteen to twenty. Three of my sisters had civil ceremony weddings because they couldn't afford a traditional one: the bride and groom wore suits and the reception was held at someone's home. The one exception to this was when my brother Jack married Merina. They had a fancy wedding in a church; Merina wore a white wedding gown; and the reception was held in a hall. Merina was so fashionable I thought she was a movie star. Oh how I loved those beautiful, fancy, frilly bridesmaids dresses.

My sister, Shirley Ann, married a black man from Jamaica, which in the early 50's was considered unacceptable in many social circles, but not in our house. My mom embraced cultural differences and saw diversity as an opportunity to learn more about others.

My home and family life was radically different than that of my best friend, JoAnne, who lived two doors down. She had a mom, a dad, a sister named Beth, and one of her grandmothers lived with them in their house. Her mom was so young and beautiful, and wore perfume and suits to go to work.

When I was six, I liked playing at Jo-Anne's house and secretly wished I could live with them. Her daddy was so much fun: he'd tuck her into bed at night and put little surprises under the pillow for her to find in the morning, and on Sundays, Jo-Anne would snuggle into his lap to watch Walt Disney on TV. Sometimes, he took her on holidays to motels with swimming pools.

JoAnne willingly shared her family, toys, and even invited me to accompany her when she visited her rich grandmother, who lived in a big house near Casa Loma in the heart of downtown Toronto. On Saturdays, her Grandma took her and her sister, Beth, for lunch at Laura Secord's Restaurant on Yonge St., or went shopping for a special dress at Eaton's Department Store.

JoAnne and Beth shared a bedroom with two single beds with matching bed spreads and curtains. In the closet, their shoes were lined up in a row, and their clothes were sorted and hung on hangers. It seemed they had lots of toys and her favorite doll even had its own crib and doll chair.

Despite not having these things in my life, I nevertheless was happy in the knowledge my Mom loved me to bits. I think the reason she loved me so much was because I didn't have a dad – it was as if she hugged and snuggled me twice as much – once for her and once for the dad who wasn't there. We did everything together: baked bread; sang songs along with the radio; danced in the living room that was emblazoned with huge green and red flowered wall paper; and slept together in the afternoons on a tiny sofa. We attended a variety of churches on Sundays. Kids loved to visit our house because Mama-mia would give them little boxes of raisins, or freshly baked small knotted buns.

Before I started school at the age of six, I went everywhere with my Mom – even on house-cleaning jobs. On a daily basis, we went on errands around the neighborhood, and I noticed how everyone loved to chat with her. Sometimes, as an impatient child, I hoped we didn't meet anyone so I didn't have to wait while she visited. There were always unexpected stops with many cups of tea shared in the kitchens of her friends, and occasionally, those visits resulted in a fresh box of crayons or a brand new coloring book (provided by the host,) while the adults shared stories about their respective lives. Most of my Mom's friends were her age, and I didn't realize how old she was until a school mate asked if my Mom was my grandmother!

My memory of those early years is now somewhat hazy and glossed over by thoughts of the happy times. We learned early in life to be frugal and to stretch each dollar. So unaware was I of my mother's struggle to provide for her family, I never had the sense that we were poor. This was a time when hand-me-down clothes were gifts, and sharing a bed with siblings and our mom was the norm. One of my fondest memories is when Ray, my mom, and I (I'd be about seven) would lie in bed, munch on popcorn, and listen to "The Creaking Door," or "Amos and Andy" on the radio.

When my siblings moved out of the house, my mother turned it into a boarding house for newly arrived Italian immigrant workers. They called her "Mama-mia." The men slept upstairs in the bedrooms, while mom, Ray and I slept in the front room on the main floor. I have many fond memories of suppers served on the table in our compact living room with three or four big men enjoying my mother's hearty home cooked bread, soups and stews. Even the language barrier didn't dampen the joyful and exuberant exchange between the men and my mom. She was happiest when she served up a meal that elicited delightful "ooos and ahhs."

I can only remember one occasion when my Mom got upset with me and sent me to bed after a good scolding. I'd gone down to the other end of the street to play with the kids from the family with eleven children, and that was out of bounds. However, I do vividly recall the bickering that went on between my siblings and my mother. The yelling made me fearful and I often clutched onto my mother's side for protection or hid under the covers to block out the shouting. I later understood this was the result of living in a crowded house under financial stress and the crutch of alcohol dependence. My mother enjoyed her beer and the escape it brought from the hardships she endured.

My mom's difficult life took a toll on her appearance and health. She didn't like doctors, and when sick, treated herself with home-made remedies. Sadly, her life came to a sudden end when she was fifty six. She had colon cancer, and didn't seek help until she could no longer bear the pain. The last recollection I have of her is when she was on a stretcher being wheeled into an ambulance and I'm waving goodbye. She died that night during emergency surgery.

Needless to say, for a child who'd slept almost every night of their life with their mother, I was inconsolable. From that day forward, my brother Ray and I were cared for by our sisters, Leah and her husband Frank, Judy and her husband Bob, and our oldest sister, Roxie and her husband Byron. Even though each sister had their own family, they embraced and nurtured us through some trying times, sacrificing and struggling to make ends meet with no financial assistance from an absent father.

## My Search for Meaning

The pain of my mother's early death and the sense of abandonment I felt, as a result of losing both parents, haunted me throughout my teen and young adult years. For most of this period, I allowed thoughts from childhood experiences, conditioning and family issues to filter my responses to life's opportunities. Those thoughts made me fearful. I worried about "worst case scenarios," and "what ifs," always trying to have a plan in mind should events happen beyond my control.

With counseling and the reading of "Self Help" books, I began to look at life in a different way and to get a better understanding of the "human condition." With this new knowledge, I learned to choose alternative ways to behave in response to situations and experiences. I learned to acknowledge my feelings, and realized I didn't have to react, or act based on conditioned past experiences. This was a liberating idea enabling me to be present and mindful of what was happening in the now. I could remain open to experience a spontaneous response to whatever happened, and to let go of feeling fearful in anticipation of what might happen.

My early year's experiences prepared me to accept life's uncertainties, and to foster a willingness to adapt and change. Over time, I shed my need to control outcomes. Remembering my mother's "acceptance of what is" attitude, I wanted to emulate her care and concern for others. Her rule was the Golden Rule: "Do unto others as you would have them do unto you." She taught me to keep my heart open to give and receive love; to do my best at all times; and to treat everyone the same. Her words, "No matter how poor you are, you always have something to share, even if it's only a smile," have stayed with me all my life. She instilled in me a need to be self-reliant, and I can still recall her saying, "If you want something, use your brain to think and your two hands at the end of your arms to help get what you need." Knowingly, I think she was preparing me for life after her death.

## The Middle Years

I had a powerful thirst for learning, which resulted in several diplomas in the social sciences field. I was a seeker on a life journey of self-discovery: of examining the connection between mind, body and spirit. What I knew then, and still believe to this day, is I wanted to be of service for some-thing/someone outside of myself. Sharing what I have to improve the lives of others is what fulfills me. I believe everyone is born with a loving heart, and it's within their learned experience how, where and when they express their compassion for others.

I married when I was twenty-five and gave birth to two daughters, eighteen months apart. Giving birth was the most profound experience that had happened in my life, and I was overwhelmed with a love I'd never felt before. I could now understand why a parent would sacrifice so much for their child's well-being. The responsibility that came with parenthood was huge. As parents, we were the guiding light in helping mold each precious bundle of possibilities and potential into an adult. What I wanted most was for my girls to know themselves; to love and care for others, including all life species; and to become confident, respectful and compassionate adults. I'm proud to say they've done that, each in their own way, and that gives me peace and joy.

## My Personal Statement of Fulfillment

Throughout the years of raising my family and building a career, I often thought about the purpose and meaning to life. What helped greatly in this quest was when I took the time to be still, to turn off the noise and busyness of everyday life, and to create a space in my mind where I could think clearly. I asked myself what did I really, truly need to be happy, and what would I walk through fire to have in my life? I reflected upon the times when I was the happiest, the most contented, and the most excited about life, and came up with a "Personal Statement of Fulfillment."

It read: "I will consciously and constructively seek opportunities that provide connection, contribution, creativity, intimacy, reflection and adventure. I will approach life with integrity, tolerance, patience, humility and remain congruent in my interactions with everyone."

What I realized upon reflection was that my happiness was to be found in the experiences I'd chosen in life. When I was in nature; whether in a park, forest, lake, beach, or in my garden, I was in a place of reflection. Times when I met and conversed with people and actively listened to their stories and learned about their life's journey, I was connecting. Times when I was involved in activities, such as reading, writing, dancing, listening to music, or attending cultural performances, I was creative. Times when I helped someone solve a problem, or joined others in developing new ideas, or in creating solutions, I was contributing. Times when I shared confidences, dreams, laughter, tears and hugs with family and friends, I was being intimate.

I made a conscious decision to stay true to my "Statement of Fulfillment" for the rest of my life, and to live life fully each and every day. I had an unwavering belief in myself and a trust in the Universe to provide me with possibilities to consider that met my "Personal Statement of Fulfillment."

## What Comes After a Successful Career?

There comes a time in life when we ask ourselves, "What am I going to do after retirement?" What comes after raising a family, building a career, and exiting the work force? I began to seriously think about this around my fiftieth birthday, and came up with a list of ideas I thought would enhance my retirement at sixty five. This included books I'd like to read; learning a new language; taking courses in Mediation and Yoga; and Teaching English as a second language. I knew for sure I'd volunteer for one good cause or another, but the big dream, the thing I wanted most, was to travel. When I was twenty, I went to Europe with two friends and visited eight countries in seven weeks. This adventure ignited a life long thirst to explore and experience different cultures and countries.

The big question for me then was how could I afford to travel, given I'd be living off my Canada Pension and Old Age Security? As I'd always worked in the social services sector, there was no Union or Corporate pension or "golden handshake" waiting for me when I finally walked out the door.

In speaking with others about their retirement plans, there seemed to be a universal predilection with money, based on the premise "you can never have enough." Many were convinced they had to postpone their retirement to accumulate enough to sustain their lifestyle! I wasn't so sure this was a sound plan for me. My concern was on how happy I'd be rather than on how much money I'd have: happiness versus wealth. I'd learned to live frugally and knew my future happiness wouldn't depend on assets or material possessions.

By the age of sixty one, I was at the top of my game. I'd fulfilled most of my career aspirations and was enjoying being the Executive Director of the Northwest London Resource Centre. I was single and owned a small, cozy home surrounded by great neighbors in a city that offered all my heart desired for recreation. My children were settled in their lives and lived in the same city, a reasonable distance from "Mom." Although I was single, I wasn't at all concerned about meeting "Mr. Right." I felt that if by chance he should appear and cross my path, it would be because of some shared experience, and not through referrals by friends or dating sites. My life was full: I had a great circle of friends; was content within myself; and surrounded by the love of family and friends.

I began to think how I might include travel to fulfill my need for adventure in retirement. I knew two weeks here and there in some all-inclusive resort wouldn't be sufficient; moreover, I was looking at going somewhere for an extended period of time, for say several months, perhaps working my way across a country. My biggest concern was if I could live any great distance away from my children for a prolonged period of time.

My dream to travel was fast becoming an overwhelming urge that couldn't be sated with just summer holidays, and by my 62nd birthday, I was ready to step out of my comfort zone to take an unfamiliar path into the future. I was confident with my independence and ability to choose my response to "whatever," and with my big bag of tricks to fall back on, after forty years in the social service sector, I felt "good to go." I was also in good health and willing to do whatever to take care of myself. I began to explore work opportunities in different cities, and even mused about serving coffee in Tim Horton's establishments as I worked my way across Canada from coast to coast.

## When The Student is Ready, The Teacher Will Appear

In 2008, I had the good fortune to meet a very special person who came to the Resource Center enquiring about volunteer opportunities. I remember that afternoon as if it was just yesterday, because that chance encounter changed my life going forward. His name was Alastair Henry and his charming English accent was what first peaked my interest. I first heard his voice when he was speaking with the receptionist. I rolled back my office chair to see the man behind the voice, interrupted the conversation, and invited him into my office to discuss some options.

He had a "professional" air about him and I sensed he might have the skills and experience we needed at the executive level of the Resource Center, rather than for the "After School Tutoring Program."

There was something exceptionally captivating about his adventurous spirit and his willing- ness to engage in life after retirement. I desired to know more about this man. I offered him a volunteer position on the Board of Directors and he accepted. He agreed to be the Board Treasurer, and within six months, became the Chairman when the incumbent's term was up.

As a Board member, Alastair attended many meetings at the Center, and of course, there was always time for a coffee break. He was an early retiree and shared with me his stories of living and working in Canada's far north, and his International Volunteering work in Asia with Voluntary Services Overseas (VSO). He also spoke to me about two Canadian volunteer sending agencies: Cuso and CESO.

As he delved deeper into his experiences abroad, and spoke about the rewarding nature of working in developing countries, my imagination and excitement soared, and I realized this was what I wanted to do.

In January of 2009, I announced my intention to resign as of June 30. In the meantime, I applied to become a Volunteer with Cuso. They invited me to attend their Assessment Day in Ottawa, which I did, and they subsequently accepted me as a volunteer and put me into their active volunteer data base. I advised them I'd be ready to leave on a placement in January of 2010. Once again, as I'd experienced on many previous occasions, the Universe had sent me exactly what I needed and when I needed it." The old adage, "When the student is ready, the teacher will appear," now had so much more relevance for me. And then I realized that the day Alastair walked into the Resource Center was the day my dream to travel began to manifest itself.

The big question was what to do about my stuff? I had a house, a car, furniture and all that came with a typical north-American lifestyle. I could of course rent out my house, put my car up on blocks etc. etc., but did I want to carry the weight of that responsibility into this new chapter of my life? I was a minimalist and many times wondered about my personal "carbon footprint." I took a step back, looked at the stuff I'd accumulated over the years, and asked myself what was my attachment to it and what value did it give me? Did it define who I was? Did it increase my enjoyment of life? Did it provide me with a sense of belonging to a class of society who could afford such things? Did it feed my ego to say I owned this, that and the other? I decided to shed the responsibilities of ownership of all material things, including my home, and officially declare myself homeless!

Although apprehensive about giving up a great job, selling my home, and letting go of "my stuff," my children supported my desire to follow my heart in flight. I took this opportunity to think about and prepare for the final stage of life. I prepared my will, had a conversation with my daughters about "end of life wishes," and gave them the items I intended on "leaving" them anyways. Letting go of my "stuff" was a great relief, and I experienced a soporific sense of freedom. I was now ready to live outside my comfort zone and experience a new way of being. I wanted to avoid any future regret for what might have been had I not said yes to my dreams.

When I was in my forties, I had a one-on-one session with a respected and well known psychic about financial security. I will always remember his weird response, because it surprised me so at the time. He said that although I might be homeless one day, I'd never be a "bag lady." How prophetic was that?

As I busied myself with downsizing, shedding and letting go, I recalled his prediction and smiled. I realized I had in fact become homeless, yet I felt rich with my intention to honor my dream.

I reviewed my "Personal Statement of Fulfillment," and noted that nothing in it alluded to owning things! The statement reflected the values and experiences that fed my soul, and this affirmation about my decision to shed my possessions, strengthened my resolve and belief in myself.

After selling my home and contents, tying up all financial matters, completing Cuso's pre-departure requirements and reassuring family and friends I'd be safe, I was ready to embark on the next chapter of my life. The only regret I had was in not doing it earlier.

I believe the Universe continuously presents us with opportunities, and no matter what our response is at the time, it's always the right one. The path we choose leads to the next opportunity and so on and so forth. We are all on a journey of self-discovery, whether we realize it or not, and learning about ourselves is the most powerful thing we can do in our quest for personal fulfillment and happiness.

## A New Beginning

Alastair and I had made a strong connection during our work together, and after he retired from the Board, we kept in contact. Our friendship deepened as we shared family histories, explored common philosophies and discussed the items on our "bucket lists." Over the summer and fall of 2009, our friendship blossomed into love, and we knew we'd each found our special life mate: a person with whom we shared so much in common, someone who lived "in the moment," accepted unconditionally "what was," could let go of expectations, and live with compassion and respect for all life species. On another level, we were like teenagers experiencing for the first time an emotional shift from having a crush on someone to a deeper sense of intimacy. Rekindling the flames of passion and romance was thrilling and energizing. There was a maturity about our feelings and expectations which allowed us to more fully understand and savor what was happening. Although Alastair was scheduled to leave for Indonesia, we both had an intuitive sense our lives would intertwine in some way at some point because we felt it was our destiny.

Alastair was disappointed when his expected departure for Indonesia was postponed several times. In the fall, he requested Cuso to take him off that placement and send him other prospective matches, because he was ready and eager to head out without further delay. In the meantime, Cuso offered me several placements, which I passed on. They weren't great matches for my skills and experience.

In early November, Cuso sent me details about an "Organizational Development Advisor" position for Youth Opportunities Unlimited (YOU) in Kingston, Jamaica. It was a good match: one where I could use my skills and experience to help make a difference, and so I applied. Within a week, Cuso confirmed YOU had accepted me.

About this time, Alastair was also offered a placement in Kingston, Jamaica as an Organizational Development Advisor with the Dispute Resolution Foundation (DRF). He applied for the job and they accepted him. We couldn't believe our good fortune.

When I decided to begin a new life chapter as an inter-national volunteer, I had no idea of what would happen, where I'd go, what I'd do, who I'd meet, or what experiences I'd encounter. I trusted in the Universe and my inner intuition spoke to me to confirm this was right for me

Shortly before our departure, Alastair and I surprised family and friends by announcing we were going to work and live together in Kingston, Jamaica. They were mostly gob smacked, but, although apprehensive, since most family members hadn't met either Alastair or I, they all wished us well. In January of 2010, we flew to Jamaica to begin our one year placements as Cuso volunteers.

CHAPTER 3

# Our Journey

## Jamaica
(by Candas)

Once the honeymoon period of working in a distant land wears off, and volunteers settle down to a routine life, it's not unusual for homesickness to creep in. Many people get depressed, quit and go home. Maintaining consistent contact with family and friends through e-mails, telephone and SKYPE, helps cushion the loss of being away from loved ones, and keeps you in touch with what's going on in their lives.

Another helpful activity is to write or blog about your experiences in detail and share them with friends and family through various social media. It can be a lot of fun for both the writer and readers, and can lead to larger projects, such as the production of this book.

The following writings are based on Candas's journals, reflecting our work and life during our volunteer placements in Jamaica and Guyana.

## Paradise

On January 10th, 2010, after a four hour flight from Toronto, we arrived in Kingston at 10:00 p.m. and were greeted by the ever smiling Mr. Mason (Neville), the Cuso taxi driver. He drove us to the Indies Hotel, which was to be our home until we leased an apartment.

The Indies was small, clean and comfortable, but no sooner had we dropped our bags, we learned there was a problem: a water problem. There was a shortage of it! This meant no toilet flushing – no matter what you "did." Sponge showers were possible, but only if you happened to be the lucky one to catch a trickle of water from the shower head. Kingston was in the midst of a drought, brought on by little rain in the fall. The situation was also compounded by an infrastructure breakdown of the water dams and mains, due in large part to the economic and political environment in the country.

We spent the first week learning how to navigate our way around the city by bus and taxi; becoming familiar with the currency and prices of everyday items; acquiring basic items, such as phones; as well as generally adjusting to the Jamaican life and climate.

On Friday, we went to the Cuso office to meet Tarik, the Program Manager, and Kerrie, the Administrative Assistant. We also met Joanne (Jo), another Canadian volunteer, who arranged for us to meet some other Canadian volunteers at Sweet Woods, a jerk chicken place for dinner. As we savored our first taste of jerk chicken and Red Stripe Beer, Jo, Anne, Mark, Laura and Jackie, who all arrived three months earlier in October, freely gave us advice on how to stay safe while exploring the sights, sounds and beauty of the island.

## Market Day

On Saturday morning, Mark escorted us to Coronation Market in downtown Kingston in an area called "Parade." We traveled by "coaster bus," an independent local bus system competing with the public Jamaica

Transit buses for riders.

    Coaster buses were smaller, seating about forty people in tight rows of five seats across. They were privately owned, usually by the driver, and the conductor's main job was to hustle prospective customers to their bus. Competition was fierce, particularly at the terminals. Travelers were physically manhandled – pulled and pushed- to take this bus instead of that. The conductors screamed, yelled and chanted the destination in Patois, creating chaos and confusion, which was a frightening experience for us newbies, Thank goodness Mark was there to guide us through the mayhem and on to the right bus.

    Coronation Market sprawled over many acres, every foot of which was intensely occupied. Vendors in stalls that were jam-packed with every product under the sun you could ever wish to buy, competed with rowdy sellers promenading around with wares hanging from their limbs, draped around their necks, or balanced on their heads. Other merchants spread out tarps on roads and sidewalks, while others navigated push carts through aisles, crammed with customers bumping elbow to elbow to get what they needed. Women weren't bashful about trying on a bra – on top of their clothes! It was sweltering, congested and loud, and I felt faint.

    We strolled among the fruit and vegetable vendors and sampled slices of this and that, here and there. It was like being in candy store – so many yummy and exotic flavors to choose from.

    After the market place, we went to Moby Dick, a well reputed Indian restaurant for goat curry and beer. It was one o'clock by this time, the sun was hot overhead, and I felt weak, disoriented and dehydrated. I was also sick with God knows what flu bug I'd arrived with from Canada. Since I didn't have much appetite, Alastair and I shared a curried goat dish. Even with my degraded appetite, I still recall how delicious it was, and how refreshed the Red Stripe beer made me feel.

## A Journey Back to Health

Six weeks prior to departing for Jamaica, I cared for a three month old baby while her mother completed her PhD Dissertation. As requested by her, I got the H1N1 vaccine, which was being widely promoted at the time to all care-givers of small babies. Sadly, I had to stop my care-giving because I came down with a bad cold. I went to see my doctor and she prescribed some antibiotics and inhalers to ease my breathing. The day prior to leaving for Jamaica, I went to see her about renewing my prescription because I was still feeling sick and was concerned about my health. She didn't feel a renewal was necessary because my chest x-ray came back clear. I was to continue using inhalers and to drink lots of juices. And get lots of rest and sunshine, she added with a smile – she knew I was going to Jamaica.

I think something happens to your health when you're on a plane, for hours on end, breathing in re-circulated air. My coughing, sneezing and profuse nose-blowing was embarrassing, and I sensed everyone on the plane was looking suspiciously at me, wondering if I had the dreaded H1N1. I thought about writing a sign on my napkin reading, "I've had the H1N1 vaccine. This is just a cold," and pinning it on my shirt for all to see.

My illness worsened and the staff at the Indies Hotel was seriously concerned when they heard me hacking away. One cleaning lady asked about my health every day and provided me with tips to help with the coughing. One day, as I sat reading in the lobby, she sat down with me. She proceeded to tell me she knew I was a very special person, a person God had chosen as a disciple for doing his work. I hadn't told her about myself other than that Alastair and I would be living and working in Kingston for a year. She recited Bible verses telling of how God and his disciples walked the earth helping the poor and the sick. She said a prayer thanking God for sending me to Jamaica.

Up to this point, I thought this very kind and caring woman was just sharing her gratitude and I was okay with that. But after more Bible verses and prayers, she stood up, pulled me to my feet, and began chanting. Now I was feeling embarrassed and concerned as to how this religious ceremony might play out. My mind was racing. She reminded me of an evangelical preacher I saw on TV one day while surfing channels. I expected her to put her hand on my head, look up to the heavens, and implore, "In the name of God, HEAL, HEAL, HEAL!" And then I would faint and fall to the ground with a thud. I was so relieved when she didn't travel down that road. But she did pull something like an imaginary veil over my head and go into a trance. Her body jerked and she gave out a small cry. I was concerned she was going to keel over in a faint from the trauma of the trance. She held my hands tightly and encouraged me to sing. I shyly refused, saying I didn't sing. I felt awkward. Was this a common occurrence in the Jamaican culture? Or had I just attracted an usual person into my life? This wouldn't be the first time. For some strange reason, I'd met many unique people in my life who sometimes put me in precarious or embarrassing situations. My family's mantra when I set off on a new adventure was: "Just don't make eye contact with anyone!"

After what seemed like an eternity – remember we were standing in the lobby with people coming and going – I gently retrieved my hands from her grasp, thanked her for her concern, prayers and well wishes, and assured her I'd take good care of myself. She smiled, and said she knew I would do some good work in Jamaica!

Not feeling any better, I set off on Monday night to seek the advice of a local doctor. He advised I was indeed extremely sick with bronchitis, and the infection had spread to my ears, throat and eyes. He gave me two antibiotic injections and two fifteen minute sessions with a nebuliser. I left his offices with six prescriptions; his personal cell phone number; and instructions to call him in the morning with an update on how my night went.

With the help of medicine, and the holistic chanting and prayers of my Jamaican spiritual healer, it took about three more weeks for me to recover from what I still believe was the H1N1 flu.

## Where Is Home

We were instructed to buy the Sunday paper to look for an apartment in the New Kingston area, the safest residential district in the city for volunteers. The rent had to fit our $42,500 Jamaican dollars (about $425 Cdn per volunteer) housing budget. Because we were co-habiting, we had $850 Canadian for our accommodation, which was more than adequate.

After two frustrating hours on the phone, trying to contact news- paper listings to set up viewings, we were only able to schedule two appointments for the afternoon. With Neville's assistance, we spent the next four hours roaming the communities within the New Kingston area looking for "For Rent" signs and viewing available apartments. We were exhausted, and fast losing hope we'd find something suitable, and were almost at the point where we were prepared to take anything just to end the search, when the Universe intervened and presented us with the perfect place.

An agent for the last listing we looked at mentioned he had another apartment we might like, but it wasn't on the market yet. The owner had just finished sprucing it up to move in, but had changed his mind at the last minute, and decided to move into his parent's house. They were going to the USA for a year.

It was close to five o'clock when we viewed the two bedroom apartment at Marley Manor. It was perfect. The prospect of making this apartment our home in beautiful Jamaica for the next year was thrilling and wonderfully exciting.

Marley Manor was set in an open compound comprising about six units, each four stories high with two apartments on each level. The spacious parking lot fronted a swimming pool and rock gardens. Mango and almond trees grew profusely in the front and back of the property. The main level boasted an open kitchen, dining and living room, bedroom, bathroom and a balcony overlooking the pool. An open staircase from the living room led up to the loft level, which housed the master bedroom. It was vast. A comfy king size bed sat up high in the center of the room, and a queen size vinyl futon lined one wall. The second bathroom was bright and airy, and there was even a laundry area on this level with a washer and dryer. All floors and most walls were tiled. The whole apartment was tastefully furnished and so spotlessly clean it gleamed. We could also accommodate up to six overnight visitors, which we did on many occasions.

We moved into our home at 6 Marley Road, Apt 334, just around the corner from the Bob Marley Museum, on Tuesday, January 19th. Our first activity was to christen it sipping an ice cold Red Stripe beer on the balcony, under the shade of an immense, overhanging mango tree.

## Settling in

How great was this? A whole year to spend on a honeymoon in paradise. Prior to leaving Canada, Alastair and I had not co-habited: he lived with his daughter and I lived in an empty house, the sale of which closed ten days after I arrived in Jamaica. Although we'd known each other for about a year through my work, we were only six months into our relationship as a couple. There was still so much to learn about one another and no doubt many confidences and untold stories yet to be revealed and shared.

After fifteen years of being single, here I was setting off to live the next year of my life as a couple. One can get pretty set in their ways of being when they live alone for a long period of time, and so there was some apprehension on both our parts in how successful we'd be in managing this lifestyle transition. I'm pleased to say that Alastair's calm and undemanding nature made living together easy.

In the beginning, there were many Skype calls from home. My daughters checked in with me regularly and asked, "How are you Mom? How are you really, Mom?" They were seeking information as to how things were going in the relationship. They wanted to be sure I was being well-cared for, so far away from home, living with a man they knew little about. I jokingly said, "Stop worrying. I like Alastair's bedroom behavior." They laughed and said, "Oh Mom – too much information – we really don't need to know about your sex life." They just needed to know I was okay and I assured them I was happy and had no regrets about my decision.

Right from the start, Alastair and I lived in acceptance of each other's odd little ways, and were comfortable enough in our own imperfections to share all with one another. Surprising as it might seem, we haven't expressed a cross word or angry response, or said a hurtful thing to one another to this very day. Maybe it's to do with our age, our maturity, and the wisdom we've acquired over the years. Life is too short to waste even a minute in a negative state of mind, and this quite possibly is something experienced by other couples who begin a new relationship in their senior years.

In the first few months of our placement, we had many dinners and outings with fellow volunteers. They were amazed when we referred to ourselves as "newly-weds." Technically, it wasn't correct, for neither of us was interested in "marriage" per se, but we used this harmless expression because it was more socially acceptable than "newly-common- lawed." Many volunteers thought we'd been married for a long time.

## Our Journey

Blue Mountains

Marley Manor

Kingston Market

## Port Antonio

After our orientation, we were most fortunate to be given an additional week to settle into our lives in Jamaica before starting work. The key people: the Cuso Director and our employment Executive Directors were out of country attending a conference. We took advantage of this opportunity and hired Neville to drive us to the Ivanhoe Guest House in Port Antonio on the island's north east side for a weekend retreat.

After checking in, we eagerly set out to explore the area. It was a busy place – a non-stop stream of cars, coaster buses, taxis and people jostling to get through the town's narrow corridors – and it was noisy. There seemed to be hundreds of unemployed people everywhere, many of whom called out to attract our attention and to engage us in conversation. If they succeeded, they believed there was a good chance we'd buy something. They all had something to sell, or could provide us with whatever commodity or service we needed. Funny how everyone just happened to have a family member or a friend who had just what we were looking for!

We strolled into the Errol Flynn Marina: originally built by the actor in his heyday in the 50's to entertain his cronies who came to the island for a visit. The marina had since been purchased from his estate by the Port Antonio authorities, and redesigned as a Yacht Club with public docking facilities. The grounds were beautifully manicured and brimming with exquisite, tropical flora. Unlike the streets leading to the marina, there were few local people in the grounds and not many tourists either, which was surprising. This no doubt was the reason why there were so few locals – no tourists to sell to!

As we were departing, we met Noel, a vibrant, cheerful Rasta man. We couldn't ignore his beguiling manner and easily gave way to chatting with him, even though we knew he must be selling something. He was all about the sounds of "old school reggae music," and said he'd make us some CDs because he knew we appreciated this type of music! We met up with him the next evening and purchased two CD's. In gratitude, he whipped off his colorful woolen hat, let down his waist-length dreadlocks, and insisted we take a picture of a real Rasta man!

Later that day, we moseyed through the large local market in the heart of town. It was jam-packed with stalls selling fruits and vegetables, as well as the usual consumables one finds in every market. Every vendor offered Blue Mountain coffee at the "best price." If you showed any interest, they'd immediately drop the price and offer a special price, cautioned with "please don't tell anyone." Even construction workmen going about their trade stopped and offered to sell us a bag of Blue Mountain coffee at a "very special price!"

In the craft market, we met an artisan by the name of Rock Bottom. He was a wood carver, and judging by the newspaper clippings and articles about himself nailed to his stall, he was well known in St. Mary's parish and quite a character. He used driftwood and Jamaica's native tree, Lignum Vitae, to create remarkable masks and household items, such as bowls and figurines. His specialty though was carving naked men with abnormally large penises. You could get small, medium or large statues, but all had proportionately oversized penises.

As we ventured deeper into the market, we came across King T-shirt. He claimed to have started the T-shirt business in Port Antonio, and readily pointed to newspaper articles he, too, had nailed to his stall. He boasted of his age, seventy eight, and encouraged us to ask anyone about his history in the market.

Most vendors had a story to tell and this was how they attracted and engaged potential customers. I thought it was a brilliant idea because it was initially non-threatening, and actually quite pleasurable. Although we didn't buy – we seldom did – we chatted with many vendors, and were interested in their life stories. We sincerely enjoyed their company.

There were noticeably few tourists in the market. Discussions with locals indicated the Jamaican Tourist Industry had tanked with the 2008 global recession and had not recovered. Even the number of cruise ships coming into the harbor was substantially less than before the recession.

On our way out of the market, we met Captain Rebo, who offered his services as a raft Captain on the Rio Grande River. He proudly pulled out his license to prove he was a skilled Captain, and assured us with him, we'd be safe. We negotiated what we considered a fair price and agreed to meet him the following morning at 8:30 a.m. at an appointed location.

The next day we took a taxi with Rebo to get to the starting point on the Rio Grande River.

The bamboo raft was about ten feet long by three feet wide, with one low seating bench for two, located close to the back of the craft. Rebo stood at the helm and navigated the raft with a long pole, skillfully maneuvering between shoals, rocks and currents, all the while sharing detailed information about the river, terrain, farms, flora, fauna and himself.

At one spot in a bend on the river, he navigated to the shore line and took us under a bridge, formed by huge overhanging rocks. That was where, according to legend, Errol Flynn took his ladies for a little romance. Rebo insisted he take a picture of us embracing at Errol's love haven – as if we needed encouragement!

Midway through the journey, he docked the raft at the riverbank and encouraged us to swim in the cool fresh water. We didn't bring our swim wear. Nor did we bring big appetites, and that was unfortunate. A local woman had prepared and laid out on a picnic table, right there on the river bank, a huge traditional breakfast of rice, peas, callaloo, dump-lings and chicken. Rebo said he told us about the swimming and picnicking, and maybe he did, but neither of us recalled that part of the conversation. Perhaps we missed it because of his thick Patois. We felt badly for the woman who had gone to so much trouble: hauling baskets of food and boxes of cooking utensils down a steep hillside from her farm. However, we were soon relieved when another raft came round the bend with two hungry passengers.

The trip continued with us snapping pictures of the breathtaking beauty encasing the river on all sides. The abundant lush green vegetation; the acres of bamboo swaying gently in the warm breeze blowing across the hillside; the ever changing shades of blue reflected off the water, and the magnificence of the Blue Mountains in the distance, silhouetted against the clear blue sky, were awesome. The warm sun on my shoulders put me in a meditative state. I breathed in the tranquility of this special place and imagined the restful energy coursing through my mind, body and soul. I was in a space that was truly in tune with my personal and spiritual essence of peace – being one with nature.

Unfortunately, close to the end of our journey, I was shocked back into reality by witnessing how a society destroys its natural environment. There on the river bank, were two huge, ugly machines, groaning and spluttering with profuse sounds of engines hard at work. A bulldozer and dump truck were stealing sand and stone from the riverbed, all the while spewing oil and gas into the pristine fresh water and air. We were told these resources were scarce in this rain forest area! But, the sand and stone were also badly needed for the construction industry, as well as sand for the all-inclusive beach resorts! It was an absolute crime in my eyes – corporate greed stealing natural resources for profit with little concern for environmental impacts! But, as is sometimes the case, there was another side to the story: poverty and desperation caused people in need to go to whatever lengths they had to secure an income for their families.

Rebo said it would be good if we, as tourists, reported what we'd seen to the authorities. But we intuitively knew, as did Rebo, the sand and stone thieves would either buy off the police or just pay the fine. Either way, they'd be back within a week continuing their efforts to make a living. Given the reality of day to day life in Jamaica, I now acknowledged I no longer wanted to wear the environmental activist's hat – ready to report and bear witness to the wrong-doings going on in the world. I had to let go. I'd come to Jamaica to help, and sensed there'd be other ways and means for me to do this once I was in my placement.

We ended our trip at the river mouth where it flowed into the ocean. Gratitude filled my heart for having spent the day on the Rio Grande River with Rebo at the helm and Alastair by my side. For the last six hours, I'd been cocooned in a tropical paradise and surrounded by nature in all its splendor and glory.

## The Well Is Dry

The Taino, the indigenous people of Jamaica, called their country Xaymaca, meaning "Land of Wood and Water." Jamaica is renowned for its mountains, rain forests and fresh water streams and rivers. The drinking water is pure and clean and there are no concerns about waterborne pests to make visitors sick. So, why was there no water in our tap at home?

The Kingston drought wasn't only persisting, but was getting worse by the week. Everyone told us they'd never ever experienced anything like it in the past.

Each apartment at Marley Manor had a separate back-up water storage tank, in case, for whatever reason, the city water wasn't flowing, such as was the present situation. Our tank was on the roof, three stories up from ground level. Water should have flowed into our taps when we turned on the tank switch but it didn't. We contacted the landlord, who arranged for a handy-man to troubleshoot and fix the problem. While we waited for him to show up, we purchased water in two gallon jugs from the grocery store.

We woke up each morning around 6.00 a.m., and went out to the balcony with our first coffee to watch the sun rise. On about the second or third morning, I noticed an increased level of human activity in the parking lot – people were coming and going across the grounds with huge water containers. On our way to work, we inquired at the guard house about the activity, and learned the City was turning the water on to our complex for one hour every morning and one hour every evening. But none made into our tap or our neighbors taps! It seemed to only be available in the parking lot taps. So the following morning we joined our neighbors (many of us in pajamas) in the yard, and lined up with a supply of two liter jugs at the garden tap by the guard house. This brought a whole new meaning to the term, "hanging out at the water cooler." Actually, it was a fun way to meet neighbors, many of whom we otherwise might never have met.

The handy-man showed up six days later, tinkered with the tank, and announced he'd fixed the problem. We ordered water to fill the tank, but as water deliveries were in hot demand, they said it might take up to a week to get a delivery. Four days later, on a Sunday night at 11:30 p.m., long after we'd gone to bed, the truck showed up. We got up and let in the water guy with his tool bag. He climbed the stairs and went through a door off the bedroom in the loft level to access the outside second floor landing. He placed a ladder on the wall to the roof and climbed up. Once he'd hauled up the hose from the water truck and put one end in the tank, he gave the okay to the driver to start pumping. The generator kicked in, and to our astonishment, water spurted out all over the place. The hose must have had twenty or thirty big holes in it. It was leaking like a sieve. We doubted if any water could make it up the three floors and into the tank.

Neighbors soon caught on and came out with buckets to catch the fountains of water streaming out of the holes. We joined them with our supply of jugs and had a good laugh. When we inquired from the guy on the roof about how much water was actually going in the tank, all he said was, "no problem man!" After a good half hour of pumping, they shut off the pump, collected their things and left. Our neighbors returned to their homes with at least a couple of day's supply of free water. Imagine our shock the next morning when we looked at the delivery slip and the company's name read "Cesspool Cleaners!" Obviously anyone with a truck was now in the water delivery business.

We had new concerns – should we even use the water and what about the tank itself – should we get that sterilized? Had it been contaminated? Could we ever use it again? Well, as it turned out, we had nothing to worry about because no water had made it into the tank. The well was dry. Our landlord was shocked and concerned when we told him we still didn't have any water in the tank. He didn't know what was going on. He arranged for more plumber visits, all of which occurred during the day, requiring one of us to stay home from work. We were still in the orientation period in our workplaces, but fortunately our bosses and the Cuso Program Manager were supportive and sympathetic to our plight. This water situation had become a national crisis.

The water tank problems continued for many weeks. Even though we had water delivered several times, none ever stayed in the tank. We laughed at our dilemma and were pleased with how quickly we'd adjusted to accepting this situation. We knew the problem would eventually get resolved, and in the meantime, we just had to be patient and hang in there. We joined our neighbors in the search for readily available water, but this time it was at poolside. We scooped up buckets of pool water to flush the toilets in the bathroom. The volume of water was of course finite and shrank with each passing day. It would soon run out, and then what would we do? We began to conserve and re-use water wherever we could. We captured all grey water from cooking and dish washing and used it to flush the toilets.

When water was available in the parking lot tap, we continuously re-filled eighteen plastic two liter jugs. We boiled the water and ran it through a Brita filter to convert it into drinking water. To do laundry, we got up at 5 a.m. and went to the laundry room to ensure we were first in line when the water was turned on. We usually managed to get one load washed and the rest we washed by hand on an "as needed basis."

Our landlord sought a second opinion, or was that a fifth or sixth? He brought in a different plumber to look at the situation. It didn't take him long to diagnose the root cause – a switch on the tank had been incorrectly installed. It must never have worked. All the water going into the tank flowed straight through and out into the City's water lines! With this fix and the purchase of water from a trusted provider, we happily soldiered on for the balance of the drought, which turned out to be three months. Suffice to say, we'll never again take water in the tap for granted.

## The Work Begins

My placement at Youth Opportunities Unlimited (YOU) was a good match for my skills and experience in youth based programs, community development and fund development. After a number of weeks of settling in and getting ourselves set up to live in Jamaica, I was eager to get started in my placement.

YOU was a charitable Jamaican organization that provided positive interventions for in-school adolescents. The organization offered a variety of vital support programs designed to help students complete their post-primary education and pursue higher education, employment, or skills training. The support programs included: One to One Mentoring; Group Mentoring; Mentoring Consultancy; Adolescent Workshops; Cultural Arts; Remedial Enrichment; Peer Counseling; Leadership Training; Workplace Orientation; Summer Development; Parenting

Education & Support; Sport Programs; Motivational 'Power Talks'; Capacity Strengthening for Teachers; Counseling; Advocacy; and a Youth Integrated Services Centre.

My original work-plan was to assess the organization's programs, develop and implement fund-raising strategies and write Staff Training Manuals. But to do this, I needed to better understand the organization and where it was at. I surveyed the Board of Directors, Staff, Consultants and Stakeholders, and used their feedback to prepare a Stakeholder's Analysis. The results showed that the most critical issues were: securing sustainable funding; strengthening the capacity of the Board of Directors; and developing a Five Year Strategic Plan. Cuso sent in a second volunteer to work solely with the Board, while I concentrated on the funding aspects.

With information gained from the survey, I developed a "Directives for Fund Development" plan, which listed steps to help ensure the organization's financial sustainability. I also developed, planned and facilitated fundraising initiatives, such as grant proposals, direct mail solicitation letters, a raffle and volunteer community outreach activities. Many unforeseen situations arose impeding my progress and creating a high frustration level. More than once I thought about "throwing in the towel," but I reconsidered, and decided to persevere. Instead, I focused on doing my best with what I had. And I'm pleased I did, because I now realize I benefited greatly in terms of personal growth.

As my understanding grew of how, when, and why things were the way they were, so did my ability to remain open and flexible to handling challenging situations. Resources were scarce, and simply not having a pen or paper for the printer, or the use of a telephone, or a working computer, could easily become a source of frustration. Sometimes the funding didn't come through to cover payroll, yet employees still came to work because they were passionate about what they did. Delays in getting paid caused them and their families' great hardship, but they always seemed to muddle through, and their ability to take things in their stride impressed me greatly. I too went with the flow, never knowing what resources might or might not be available that day, or who may not come to work or be absent from the workplace for weeks on end. I approached each day with the conviction that I'd do my best, no matter what challenges may arise.

My work content was diverse and interesting, and I had to wear many caps I never knew were in my closet. One day, I could be meeting a local parish Mayor about the need for community input for his intended building of a resource center; the next day could see me interviewing potential volunteers; and the day after that, teaching telephone etiquette to forty summer students. Sometimes, I was the receptionist: answering both the door and the telephone; or the lunch maker: making sandwiches for the summer students; or the writer: composing a letter to Michelle Obama asking her to be the Key note speaker at the first annual Caribbean Mentoring Conference. Unfortunately, the Conference plans had to be put on hold as the funding issue became critical and needed to be addressed. I worked from the premise I wasn't an expert, merely someone with experience to share and to support the already accomplished staff to do what they did best.

## Alastair's Work at the Dispute Resolution Foundation

Jamaica has one of the world's highest per-capita murder rates, but I'm pleased to report that it's fallen in recent years due to increased police patrols, curfews and more effective anti-gang activities. Some areas, particularly cities such as Kingston, Montego Bay and Spanish Town, experience higher levels of crime and violence. These crimes typically involve attacks by Jamaicans on Jamaicans, not on tourists, and revolve around drugs, gangs, politics, poverty, and revenge.

The Jamaican Justice system was so swamped with the number of civil cases on the books to be heard by so few judges that in 1994, the Mediation Council of Jamaica was incorporated as the Dispute Resolution Foundation (DRF), as a way to resolve disputes and decrease the case load. The situation at that time was so dire that it was taking fourteen years from the date of filing a case to get to trial. The delays many times resulted in minor disputes festering, and becoming much larger issues and evolving into criminal cases. DRF developed a methodology to train people as mediators, and over the years, expanded its training services to include restorative justice, arbitration and community justice. It now had a network of trained mediators in every parish island wide, and trained a diverse group of people, such as lawyers, teachers and correctional officers, in both Jamaica and other Caribbean countries.

Mediation is a voluntary method of dispute resolution. It's confidential, quick and low cost. Mediators are neutral third parties (they don't take sides or impose solutions) who use a facilitative, communicative approach to help all parties arrive at a mutually acceptable agreement. In Jamaica, it's been highly successful, mostly because if parties don't reach an agreement, it goes back to court, and joins the back log of cases to be heard at some distant date in the future.

In 1999, Jamaica passed a law adopting optional mediation with DRF as the referral agency in civil court filings, and in 2007, made it automatic for most Supreme Court civil cases.

In 2005, DRF and UNICEF developed a youth program called "Conflict Resolution as a Solution – Creating a Safe Place for Learning," and implemented it in some schools. In 2006, they renamed it the "School Suspension Intervention" program, and it was now in a hundred schools island wide. When children get suspended from school, they must go to the DRF office and work with a trained Youth Councilor on whatever problem they're having, such as anger management. And they must continue their visits until the Councilor feels they have satisfactorily resolved their issues. It's about the offender taking responsibility for their actions, and in letting the victim have a voice.

DRF also established two Community Peace and Justice Centers in three of the most troubled neighborhoods in Jamaica: in Trench Town and Spanish Town, Kingston, and in Flankers, Montego Bay.

Alastair was helping DRF develop a five year plan to further expand and strengthen their operations across Jamaica and into selected Caribbean countries, namely Belize, Guyana, Trinidad and Tobago, Barbados and Bahamas.

He took the week-long mediation training course and was so impressed by its simplicity and effectiveness that he felt it should be taught as a life skill in every school and University in every country. And he added, because if it was, adults would handle their relationships differently, and there'd be less divorce, disagreements with neighbors, and stress in the world. And if politicians and governments used mediation to resolve their disputes, there'd be fewer wars and a more equitable balance of power in the world. He was enjoying his work and the camaraderie of staff very much.

## Pegasus

I shall remain forever grateful to Pegasus. At some point early in February, we joined other volunteers after work for a swim at a hotel in New Kingston called "Pegasus." Joanne (Jo), a twenty-something year old volunteer from British Columbia, took it upon herself to talk to the hotel manager about the plight of twelve Canadian volunteers working in Jamaica during the water drought. Jo was an exceptionally polished and savvy marketing professional in her real job back in Canada. Using her finely honed negotiating skills, she secured a free pool membership for all Cuso volunteers for one year.

It was most fortunate our work places (Alastair's and mine) were right across the road from one another, because it simplified our lives. We could travel to work together and meet up at day's end, without the risk of scheduling mishaps. We went to the Pegasus at least every other day, and if we had an unbearably hot and humid spell, we'd go every day. It was a thirty five to forty hot minute walk to the Pegasus and by the time we got there, we were usually exhausted, lathered in perspiration, and craving fluids. Our first priority was a hot, soapy shower, for not only were we sweaty from the commute, but we were also unbathed – there was no water to shower at home. Clean and refreshed, we jumped into the sparkling, cool, twenty five meter pool and swam, splashed and luxuriated in the water for the next hour. It was delicious. A second shower to rid the chlorine, and it was time to eat.

Also convenient, was that our bus stop was right outside the Pegasus and a bus for home came by every ten or fifteen minutes. We'd be home and eating supper a half hour later, if we so chose, but many times we ate out. We could have eaten at the hotel restaurant of course, but it was too pricey for us. More in line with our budget was the Pizza Hut up the road or Sweet Woods across the street. They had a wide assortment of jerk meats – chicken, pork, fish and beef – as well as festival (a traditional fried roll of dough), and Red Stripe beer, all reasonably priced, and an outdoor seating area to boot that we particularly enjoyed.

Occasionally, we lucked into the Manager's Social Night at the Pegasus. It was an evening of free food and music to encourage hotel guests and club members to mingle. We were members, albeit not paying ones, so we stayed and enjoyed the delightful free hors d'oeuvres, such as fried sweet potato, dumplings, festival, breadfruit, jerk chicken and fish, as well as a great variety of deserts, topped off, of course, with a glass of wine or beer. We enjoyed mingling too.

The opportunity to swim in the pool, laze in the sunshine, and treat ourselves to a sumptuous meal at their upscale restaurant once in a while was indeed a most welcome gift from the winged Greek horse. It cushioned the culture shock of living in a different environment, and helped us stay healthy and happy.

Most Kingston-based volunteers used the Pegasus as we did and built their weekend activities around visits to the pool. We preferred to go to the Pegasus early on the weekend and have the whole place to ourselves for the first hour. And then, as the day heated up and more people appeared poolside, we moved on to tackle our weekend chores.

## Thoroughly Grilled

Jamaica has a long history of violence. It was no surprise then, that prior to and after our departure from Canada, many people spoke to us about safety. We'd done our homework and felt well prepared, however, the continuous warnings and urgings about safety were concerning, and as it turned out, well warranted.

Soon after our arrival in Jamaica, we went to a soccer game with a group of Canadian volunteers to watch Jamaica play Canada. When we arrived at the stadium, we learned we couldn't buy tickets there: we had to get them at either a convenience store, or a gas station! How ridiculous was that? But if that's the way it was, then that's what you did. We needed thirteen tickets and it was half an hour before kick-off. Where were we going to get them? Well, as it turned out, scalping was legal in Jamaica; moreover, it was a common and acceptable way for a local person to earn a living. Many police officers were stationed by the main gate side by side with the scalpers!

After some haggling, we purchased the tickets and bought some cold beers and water from the vendors outside the gates and went into the stadium. It was necessary to pass through several security checks, requiring pat downs and examination of bags. At the final security check point and ticket turnstile, they told us liquids of any sort weren't allowed into the stadium! So why were they selling them at the entrance then? We had a choice – drink up, or toss the full containers. Being volunteers, and frugal ones at that, there was no way we were going to waste our purchases and so we drank up.

Once inside, we looked for the Canadian cheering section – for Canadian red and white T shirts, face stickers, flags and pins etc., but we couldn't spot any in our section. There was such a group, but they were in the premium seating area, under shelter from the hot sun. They were obviously not volunteers on a stipend like us.

As soon as we stopped and took up a position to watch the game, hordes of vendors selling cold beer, water and a variety of snacks, appeared and hustled us to buy. Go figure. "What a scam" was the only thought that came to mind. How could a beer purchased outside the stadium from a local vendor be any more a security risk than a beer purchased inside? Makes you wonder about the grafting going on between inside vendors and outside turnstile ticket takers, doesn't it?

Attending the game provided us with more insights into the Jamaican culture. They loved their soccer! It was a fun and exciting event, even though Canada lost. As we departed the stadium, patiently inching our way forward along with hundreds of other fans, I saw the man next to Alastair point his lit cigarette threateningly at Alastair's neck! I yelled at him, alerting Alastair to the danger and to the impending crush of human bodies pushing us towards a wall. We shoved and pushed our way out of the unfolding situation and bolted up the road to an open and lighted area to wait for our companions.

Another volunteer, Dominic, was similarly caught by the hoodlums. They swarmed and isolated him against a wall and tried to get into his pockets in search of money or his cell phone. But they'd unwittingly picked the wrong guy. Dominic was a most experienced traveler and he'd pinned his pockets with safety pins before he left the house to prevent being robbed. He screamed at the robbers, alerting the crowd something bad was taking place. The robbers immediately dispersed and Dominic joined us: a little shaken up, but grateful he was unharmed. This had been a classic close encounter with the "safety issues" everyone had warned us about. We all agreed we would be more mindful in the future about our surroundings when we were out and about in public gatherings.

It took that incident to truly make me aware of the security threat ever present in Kingston. All homes, apartments, businesses, stores, schools and churches were protected by steel grills on all doors and windows. Most buildings had barriers around the outside of the grounds consisting of cement walls, six to eight feet high, with barbed wire, spiked glass or metal spikes on top. Each driveway was secured with a tall metal gate and most had a ferocious looking dog.

For some strange reason, Jamaican dogs didn't like me. Whenever I passed by the gates, the dogs would race to the fence, excitedly jump around, gnash their teeth and bark their heads off. I was scared one would get through a hole or jump the fence and rip me to bits. I like dogs. In fact, they're one of my favorite animals. So why did these dogs find me so threatening?

The barking dog situation was also a problem at night: some wailed, some howled and some cried all night long. It disturbed my sleep and I became anxious. I know I rambled on about it for days and was grateful to Alastair for his patience in listening to me. A dog barking through the night was an invasive, unpleasant, public nuisance and totally unacceptable.

I came up with a plan. I would draw up a petition, get it signed by the other Marley Manor tenants, and give it to the police for consideration and enforcement. It was to get dog owners to take their dogs inside at night so people could enjoy restful sleeps. I thought I'd try it out and so presented the petition to an elderly neighbor living downstairs. I'd chatted with her previously about the dogs and thought she'd be supportive. She wasn't. I was shocked when she refused to sign. She said the dogs provide a sense of security and keep the neighborhood safe from bad people. They regard the incessant barking as a sign of protection and that lets them have a sound sleep! What was I thinking? Embarrassed, I slunk back to our apartment in the realization that I'd made a North American assumption about noise pollution that didn't fit Jamaican culture. It took a bit of fine tuning, but I eventually learned to fall asleep with the secured knowledge the barking dogs were taking care of us.

Our condo complex had a 24-hour security guard on the gate, as well as a night watchman, who constantly patrolled the grounds while we were asleep. There were six separate entrances to our complex, each with its own metal gate. Our apartment door had two dead-bolt locks for entry and a third sliding bolt-latch on the inside. The door to the balcony had a lock, as well as a metal grill with two sliding bolts (top and bottom), and a padlock for each bolt. Even though we were on the third floor, all windows had locked grills on the inside.

And yet, even with all this security paraphernalia so evident everywhere, and the recent security incident, I can honestly say that at no time did I feel afraid. My daily experience with the local people reflected the best in humanity. They were kind, concerned, and generous, and I thoroughly enjoyed interacting with them. I focused on the abundance of good will surrounding me every day and that was sufficient to allay all fears.

## A Banking Dilemma

The standard pay day for everyone in Jamaica was either the fifteenth of the month or month end, and everybody was paid by check. The only way to cash a check was to go to the bank where you have an account – no one else would cash it. Is it any wonder then, for two days a month, banks were like Grand Central Station – so many people coming and going. Many times, customers had to wait outside, because the bank guard limited the number of people inside. As one customer left, the guard let in one replacement. Why wouldn't an employer, such as Cuso, pick another day to be pay day and give their employees a break? Tarik laughed when I asked him this. "It's always been this way in Jamaica," he said.

Our first choice was the Scotiabank, which we used at home for our credit cards and banking. Armed with an array of paperwork – passports, immigration work permit, credit cards, banking cards, an introduction letter from our home bank (we were instructed to get this before we left for Jamaica), as well as a letter of introduction from Cuso, we set off with our first living allowance check in hand to open up a bank account.

Upon arrival, we were greeted by the Security Guard who directed us to the lady at the Information Desk. She listened to our inquiries and directed us to an assistant, who took the documents and went off to speak with an Accounts Manager. He returned ten minutes later with the news we required additional information: a letter, sent by mail directly from the Scotiabank Manager in Canada to the Scotiabank Manager at that branch in Kingston, Jamaica! Faxed or emailed letters weren't acceptable. We also needed two reference letters from persons of authority (lawyer, clergy, or police officer) in Canada who knew us personally. When we expressed concern about the impracticality and length of time it would take to arrange such letters, the assistant suggested we try another bank! We didn't want a loan, or a mortgage, or a line of credit – we just wanted to cash a simple check. And then he walked us to the door, escorted us out into the street, and pointed to the First Caribbean Bank up the road.

We got the same story there: we needed additional paperwork. However, the one thing we had going for us was this was Cuso's bank, but not the same branch. They were willing to cash our checks, but only after faxing over copies and our passports to Cuso for verification.

Finally, after waiting for over an hour, we were escorted to a teller and received our money. But, the teller said, this was an exception and they wouldn't do it again. Future checks would have to be cashed at the correct Cuso branch. To cut a long story short, we never opened a bank account during our year in Jamaica. We kept our money under the mattress.

It didn't matter what time we went to the bank on paydays, the line-ups were always horrendously long. On our third visit, a teller told us about a special concession for seniors: a separate seating area with two dedicated tellers for customers over the age of sixty. How sweet was that? This usually reduced the wait time, but not always, for the other section had many more tellers. We smiled and waved from our privileged seats in the pensioner section to the younger volunteers standing in line. But occasionally, they got to the wicket first, and rubbed it in. It was a little game we played.

## Red Light Hitchhikers

Each day started and ended on our apartment balcony. We watched the sun come up over the Blue Mountains, which seemed only a stone's throw away, and in the evenings, gazed at the beauty of the moon, sometimes riding high in the sky above the mountain peaks. So when it came time to venture out of town, we were eager to head into the Blue Mountains to see them up close. Several volunteers had recommended a guest house called Mount Edge, and that's where we decided to go.

We took a route taxi to a little town called Papine, located at the base of the Blue Mountains, where we were to look for a coaster bus to take us up to Mount Edge. Papine was where farmers came down from the mountain with their fruits and vegetables to sell to people, who came in from the city to buy. Though only a little square, bordered by narrow roads on all four sides, it had a big city energy and noise. Farmer's market stalls and street vendors of every stripe crammed three sides. On one side, Coaster buses lined up, like soldiers waiting for battle. It wasn't organized, at least not to my eyes, not in the way I was used to, in that there were no signs to say which bus went where. The buses seemed to just come in and park wherever there was room. It was chaotic.

The only way to resolve our dilemma was to find someone who understood enough English to know what we were asking, and then, hopefully, to be directed to the right bus. We maneuvered our way through the bus maze, found a man who seemed to know which bus went where, and got on the bus he indicated. There was no-one on the bus so we just sat there and waited for something to happen. Eventually, the driver showed up and confirmed he was going to Mount Edge. And then he added, "Bus come soon," a phrase used to reflect the immediacy of the coming or leaving of a bus.

As we were the first passengers, we had a choice of seats. Well, being first and having the good fortune to choose seats also had a down side which we were soon to learn about. You see, the bus only departed when there was a bum in every seat. And there was a good chance six bums would eventually be squeezed into the five available seats in each row. After an hour of sitting and baking in the hot bus with no air conditioning, with all seats now occupied (and three more people standing), the journey up the mountain began.

The road was narrow, pot holed, steep and winding. One side overlooked the cliff and the valley below, which of course got steeper with every mile traveled: the other side was rock face. There was no room for unexpected stops, or to safely pass slower moving vehicles, yet many vehicles did. Sometimes, we slowed down to drive around a rock pile that had fallen down the mountainside. Other times, we moved into the oncoming lane because our side of the road had disappeared into the valley below!

To drive on these roads required skill and mindful presence because you never knew what was coming around the next bend. Large Jamaican Transit System buses, smaller Coaster buses, as well as Route Taxis and private vehicles, all had to negotiate the treacherous twists and turns through the mountains. Buses stopped here and there to let passengers on and off, and sometimes, buses and other vehicles stopped at Jerk Chicken vendors to pick up lunch for the driver, or lunch orders placed by phone and to be dropped off on route.

Then there were the trucks: all sizes and in various states of repair and disrepair, carrying everything one could possibly imagine. Many were coming and going from the docks in Kingston to the north shore. Some had high loads, all strapped down of course, but they swayed menacingly in front of us, and more than once, I saw loads tilting precariously to one side, threatening to topple over at any moment.

Many of the cars were clunkers that wouldn't be allowed on Canadian roads, but I also saw some behemoth SUVs, bigger even than anything I'd seen in Canada. It was scary and concerning when vehicles passed one another in a space, much too close for my comfort.

School children in neat, brightly colored uniforms sauntered and skipped carefree along the nonexistent curb of the highway. Street vendors –women and men- carrying huge baskets and bags brimming full with snacks and frozen "baggy drinks" (small plastic baggies filled with a sugary juice with a straw sticking out) waited at every stop, and sometimes entered our vehicle to sell their wares. We were traveling through the very heart of rural Jamaica. There was so much to observe and savor. I resolved to stay in the moment and absorb it all.

After about an hour, the bus stopped and people began to get off. There wasn't much there: just a Jerk Chicken vendor, a small hut selling bread and a bus shelter, where a crowd of school children chatted and hung out. The last of the passengers departed the bus, and to our surprise, the driver told us we had to get off too because this was the end of the line. This was "Red Light."  We were stunned.

"You said you were going to Mount Edge," we told the driver.

"This is the stop for Mount Edge," he replied.

"Well, how do we get to Mount Edge from here then," we asked.

"Flag down a car," he curtly replied.

Mount Edge was another three miles up the road! Feeling uncertain about this next part of the journey, we sat down on a roadside wall and considered our options. It didn't take long for a group of curious school children to assemble before us and inquire where we were from, were we married, what we were doing, and what was Canada like etc. We enjoyed our little chat with them, despite the language difficulty. They understood everything we said, but we had trouble with their Patois.

After about twenty minutes, a car came along, and with little thought I flagged it down. The driver was reluctant at first to take us to Mount Edge, but when I said we'd pay, he changed his mind. With a sense of relief, we bundled into the back seat. One look around, and I instantly realized how carelessly I'd cast caution to the wind. The car's interior was stripped completely down to bare metal and the back seat was loose – it wasn't bolted to the floor.

And there was a second person in the vehicle in the front seat, a rough looking man. The driver gunned the engine and sped off at great speed up the winding, washboard road. He was angry and argued vehemently with the rough looking man. This was fast forwarding to a possible worst case scenario, as I recalled the briefings about safety and using common sense. My mind was racing and I could feel my body heating up with worry. We were so vulnerable.

It had rained a lot lately in those parts, and that's why the road was so washed out. Too bad the rain hadn't moved fifty miles south and into our taps! Was that why the coaster bus only went as far as Red Light! It was a hair-raising ride, like being on a wooden roller coaster at Canada's Wonderland. It was shaky, scary and nerve racking. I'd given up on thoughts about staying in the moment, and now resorted to prayers.

But we did arrive safely, ten minutes later. We gave the driver about $5 Cdn. for his trouble, and with a huge sense of relief, climbed out of the vehicle. "Far too much," Michael Fox, the proprietor of Mount Edge, quipped when we checked into his place and told him what we'd paid. He took us on a tour of his small, but sprawling facilities, perched on a cliff in the Blue Mountains 3,500 ft. above sea level. None of it looked sturdy or safe. But it must have been because he'd been operating it as a guest house for many years.  The main building had two private rooms with a shared bathroom, a spacious living room full of comfy couches and pillows, and a dining room. There were also three cottages, all with double beds and their own bathrooms. Another structure had bunk beds for four and an adjoining bathroom with hot and cold water. Most rooms had breathtaking views of the mountains.

Some volunteers came to Mount Edge as we did, simply for the break from hectic Kingston. Others came to hike down and along the river below; to take a Blue Mountain coffee tour; or to ramble around nearby Holywell National Park. Other local activities included bird and butterfly-spotting, and a visit to the nearby School of Vision, a Rastafarian community.

Our cabin, though tiny, was scary to look at from the outside, but once inside, it felt structurally sound. The views of the mountains above and the valleys below were stunning.

Three other guests: a young lady from the Netherlands, two young men from Australia, and a guy from Scotland, all traveling independently, had just finished their lunch when we arrived, and were getting ready to go into town with Michael for some groceries and libations (both in liquid and tobacco form as we would learn later). It was deliciously quiet after they left. We stretched out on the chaise lounges on the main house deck and spent the afternoon reading and napping in the sunshine.

Michael's partner, Mary, cooked a delicious Jamaican meal of jerk chicken, rice and collard greens for supper, accompanied with what else: Red Stripe beer. Everyone was present for supper, and afterwards, sat around and chatted.

Michael was quite the character: He was laid-back and not the least bit shy in voicing his strong opinions on a broad range of subjects. We started off on world politics and then moved on to more controversial topics, such as religion and social issues, including the legalization of cannabis in various countries of the world. And for the record, everyone was in favor of legalizing it. For desert, Michael shared his passion for old movies, particularly those starring Humphrey Bogart. What an odd topic to bring up with a group of strangers from different parts of the world, yet everyone seemed to have seen the movies he mentioned and fully participated in the discussion that ensued. We enjoyed the company and found the conversations stimulating and thought provoking.

When the room got thick with smoke, I took my leave. Alastair stayed. I went to our cabin, climbed into bed, and within seconds was fast asleep. I didn't stir once. In the morning, after sleeping soundly for nine hours, I thought about my strange slumber and concluded it must have been because of what I'd inadvertently inhaled earlier in the evening!

On weekends and statutory holidays, Alastair and I explored other areas of Jamaica. We traveled to distant parishes to observe life in small villages and stayed in local guest houses. We made an extra effort to appreciate the beauty of the flora and fauna, the new bird songs unfamiliar to our ears, and the sparkling, turquoise blue sea because we were mindful that human nature being what it is, one unfortunately quickly adapts to a new environment and begins to take everything for granted. The novelty of what's new soon wears off.

We talked about our travel experiences with our Jamaican born neighbors and co-workers, and most admitted to never having been to the places we'd visited. They loved our stories and adventures though, marveled at our photos, and admired our courage in taking buses and taxis: a transportation mode they'd never dare take. They were proud of their country and pleased we were interested in visiting areas outside of the usual tourist attractions in Negril, Montego Bay and Ocho Rios.

We didn't spend much money on travel. Using local buses and route taxis and staying in "out of the way" guest homes was quite economical for us, but even that little amount of money would be too much for most Jamaicans. They didn't have any money left for touring after paying rent, utilities, groceries, car and children's school fees etc.

## Port Clarence

Although Kingston is located on the waterfront, it doesn't have a clean, accessible beach. Mind boggling quantities of plastic bottles and bags, Styrofoam, and other materials get washed up on shore with the tides, and because there was no formal recycling or conservation programs in place, the garbage just accumulated year after year. This isn't a new problem. Every beach around the world has to be cleaned up by someone every day to keep it litter free and usable. We didn't think about this before.

One such community that did keep its beach pristine was Port Clarence, just outside Portmore, about an hour west of Kingston. Volunteers, who'd been in country longer than us, raved about it. We decided to check it out for ourselves, and went there on the following Wednesday because it was Ash Wednesday: a religious national holiday in Jamaica.

We left home around 8:00 a.m., flagged down a route taxi to take us to Half Way Tree, where we boarded a city bus to downtown Kingston, and then got on a coaster bus to Port Clarence. The whole journey took no more than an hour.

City bus rides on Sundays and religious holidays were interesting and most entertaining for me because they provided an unusual insight into Jamaican culture. As most passengers were on their way to church, they wore their finest clothes. Men and boys wore shiny black shoes, long dress pants and white shirts, many of which were heavily starched and ironed, and matching ties. Women and girls wore high heeled shoes, over-sized decorative hats and matching purses. Their billowing summer cotton skirts went down to their ankles. Hair on both sexes was well groomed and styled. Men and boys hair was usually slicked down or close shaven, whereas the hair on the females was a much more elaborate affair. Some silky hair flowed over shoulders and spilled down backs; some was tightly braided in a zig-zag fashion or in straight rows; some was twisted and tied or piled on top of the head. Many little girls had bobbles, clips and ribbons in every color worked into their creations.

Topping off this pleasing Sunday visual were the many hands clutching well-worn Bibles. These Books were such an essential Sunday fashion accessory that children looked under-dressed if they didn't have one in their hands. Another aspect I found interesting was the heavy air of solemnity that filled the bus on Sundays. Was that because of the many Bibles on the bus? Were children less inclined to misbehave when they had a Bible in their hand?

Music is big in Jamaica. Wherever you go, even on a bus, I can assure you'll hear music, and it will be loud, ear-splitting loud mostly. Bus drivers decide what to play. On Sundays and religious holidays, it was Gospel music. It pulsated through the bus, prompting passengers to sing out in full voice. Occasionally, when there was no music (the bus driver's choice), an individual would take it upon themselves to stand in the aisle, read verses from the Bible, and deliver a sermon. Church-goers in the crowd would add, "Amen." I appreciated the gracefulness and vitality with which Jamaicans spontaneously embraced their religion; the ease and joy they expressed in singing; and their confidence when they recited a speech, a reading or a prayer. Jamaican youth sure seemed to have a lot of confidence and that both surprised and pleased me. I had a feeling that these young people would be able to make their way in the world, even if they didn't get a full education.

The entrance fee to the Port Clarence beach was about $1.50 Cdn. What a great deal that was – such a small amount for a big piece of heaven. The privately owned, well maintained, black sand beach far exceeded our expectations. The water was a post-card shade of turquoise; the sand was black, hot and soft; the sky was clear and blue; and a salty ocean scent floated on the warm and gentle breeze as it blew in with the waves, and rustled the leaves of the almond trees.

To our surprise, we were the only ones on the beach. We selected a spot close to the showers, wash rooms and fish hut, and spread our towels under an almond tree. Within minutes, Jacob, who rented lounge chairs, visited us. As I said earlier, Jamaicans were passionate about their country and history, and felt compelled to speak about it with visitors. They also liked to know where you were from, how many children you had etc. etc. Once you got talking, they frequently digressed and began sharing personal stories, going on to tell you about their life. It was obvious Jacob enjoyed speaking with us, and we in turn enjoyed his company. Over the coming months, we got to know a lot more about him, and he about us.

This familiarity bordered on friendship although there was an embarrassingly sad sense in us (fear based western conditioning of being conned) that he had a hidden agenda, and what he really wanted was to sell us something, or to ask for our help on some financial matter. We tried not to let that thinking influence our genuine interest and willingness to open ourselves up to his kindness. For the record, he never did ask for anything except that we rent a lounge chair from him for the day.

For the first hour, we had the whole beach and sea to ourselves and it was glorious. The waves were gentle, allowing us to effortlessly lie on our backs, gaze up at the sky, and be awed by the sight of heavy pelicans soaring and diving for food.

Around eleven, other visitors arrived and populated the beach and water. Many families with beautiful black and brown skin tones played in the waves, and the shrill laughter of children filled the air. Reggae music (old school) pulsed through the trees, delighting us picnickers. As the morning progressed, we went in and out of the water to cool and refresh our bodies from the sun's intensity.

Alastair read me a short story from Robert James Waller's book of essays entitled, "Old Songs in a New Café." The writer, author of the "Bridges of Madison County," has a profound way of describing simple life experiences and people's ways of being. The story was rich in detail and Alastair's soothing voice and manner of reading lulled me into a dreamy space of contentment.

Lunch came from Tianna's Fish Hut, located about two minutes from where we'd settled on the beach. The fish was fresh, having been caught that very morning. We each had a deep-fried, meaty red snapper accompanied by two pieces of festival, a serving of sweet potato, and a Red Stripe beer. About an hour before you eat, you go to the Fish Hut and hand pick the fish you want from the dozen or so in the ice box, and place your order. What a business! It was non-stop traffic, and deservedly so, because not only was it a unique experience, but the fish was amazingly good and the price was right.

We headed home soon after lunch because dark, storm clouds began brewing overhead, and we didn't yet know how to get home. The guard at the gate house told us to walk one kilometer to the main road if we wanted to catch a bus or taxi. Fortunately for us, the rain held off and we made it to the round-about before the heavens opened. Just as the rain began, a red car slowed and the driver waved to us. It was our landlord! Imagine our disbelief at this turn of events. How serendipitous was that?

## Scary Newspaper Headlines

It was a workplace requirement for all staff to read the newspaper every day to stay current with what was happening within the youth sector in the country. Unfortunately, one was faced with the stark reality of how brutal life in Jamaica was for some people because the daily newspaper headline, in three inch bold type, reported the accumulated number of murders across the island for the year. On May 1st, the headline read, "535

Murders in 118 days." That's more than 4 people a day! And Jamaica's only a relatively small island!

This violence mostly reflected gang wars being raged in poor neighborhoods. For the most part, they were retaliation murders – Jamaicans killing Jamaicans. Each gang had a "Don," who controlled what happened in their "crib." The gangs operated protection and extortion rackets, and corruption by these gangs had, unfortunately, impacted law enforcement agencies, and influenced country politics.

During the first few weeks in May, tension and violence increased in downtown Kingston. The United States wanted to extradite Jamaica's most wanted criminal, Dudus Coke, to stand trial on drug charges.

After his father's death in 1990, "Dudus," at the age of twenty one, became leader of the gang and the Don of the Tivoli Gardens community in West Kingston. He developed community programs to help the poor and had so much local support that Jamaican police were unable to enter this neighborhood without community consent.

When the extradition papers were finally signed, all hell broke loose. The search for Dudus provoked violence among his supporters in West Kingston. The Jamaican Government declared a state of emergency, invoking Military Rule. This brought a heightened sense of alarm to the city as businesses and schools closed, workers were sent home, people rushed to stock up on food supplies, and a 6:00 p.m. curfew was instituted. Truckloads of army personnel with big guns suddenly appeared on streets and helicopters buzzed overhead as the country searched for Dudus. Beautiful and exotic Jamaica was at war with itself!

He was eventually arrested and extradited to the United States (US) in 2010. In 2011 Dudus pled guilty and was sentenced by a Federal Court in New York City to 23 years in Federal prison.

During this time of civil unrest, Cuso implemented plans to ensure the safety of their volunteers. All volunteers were sequestered in their apartments and were in daily contact with the Program Office. An emergency plan was ready in case we needed to be evacuated. As it turned out, Alastair and I, along with four other volunteers, had already left Kingston and were at Great Huts in Boston Bay on a four day holiday. We were enjoying the sand, sea and sun on the island's north east shore when the trouble in Kingston broke out. Cuso contacted us to advise the organization's position on the current situation. They instructed us to call them every day for updates, but in the meantime we were to stay put where we were. How difficult was that going to be?

On the fifth day, Mr. Mason came to take us home. He stopped at a grocery store on the way so that we could stock up on supplies. We stayed in our apartment for the remainder of the week and returned to work the following Monday. We were requested to call the Program Manager daily for updates, directed not to go into the downtown area, and to continue with the 6:00 p.m. curfew until further notice. This situation lasted another two weeks.

It may seem strange to readers that neither of us felt the least bit threatened or at risk during this tumultuous period. We trusted Cuso to keep us safe and they did. We also learned from our neighbor, Clive, that as visitors, we were safer than the average middle class Jamaican. How come? Because most Jamaicans understood the importance of tourism to the country, and if a bad incident happened involving a tourist, the worldwide media would quickly pick it up and potential tourists would cancel their Jamaican holiday. Clive also said robbers feared how badly the police would treat them if they were caught!

Once Dudus Coke was captured and sent to the USA, Jamaica settled down somewhat and life returned to normal, although violent gang incidents continued to happen, and be reported daily in the newspapers.

## Great Huts Guest House

Alastair and I visited Great Huts on Boston Bay twice during the year we worked in Jamaica, and it stands out as one of my favorite resorts of all time. Boston Bay is on the north shore of Jamaica, just east of Port Antonio. It's famous throughout Jamaica (and elsewhere in the world) as the birthplace of Jerk chicken, and claims it has the best jerk in Jamaica. That may or may not be true, but I'd agree with most people it's got the hottest jerk on the island.

Jerk Chicken is thought to be the result of African meat cooking techniques, brought to Jamaica by the slaves, combined with native Jamaican ingredients and seasonings used by the Arawak Indians. Smoking the meat for a long period of time keeps the insects away, preserves the meat longer, and imparts a strong smoky flavor to the food being jerked.

Boston Bay is easily missed when driving as it's a small bay and barely visible from the road. Great Huts was even more difficult to find. Tucked away and secluded down the end of a dirt road, we had to ask a local for directions, and it was only when we were at the front gate, looking at the small sign, that we realized we'd arrived.

Great Huts is a resort for nature lovers, who enjoy a rustic, bare bones type of accommodation. To us, it was paradise.

When we passed through the bamboo gate, we felt we'd entered the world of Robinson Crusoe. The main building was a two story structure reminiscent of a rock cave. It contained wooden benches with pillows, a desk built around a tree trunk, and an open kitchen with rock counters at which patrons sat for breakfast. The upper deck of the main building was the dining area where dinner was served nightly, and on Saturday night, a local cultural dance and song group provided the entertainment. There was also a DJ who played a variety of music, but no matter what he played it always had a strong bass beat that made your feet want to dance.

The resort was built on the outskirts of a Jamaican rain forest, on graduated levels of lava rock amidst overhanging vines and trees. The huts differed in construction and design: some were round, two story towers; some were tree houses or tents and others were clay structures with mosaic and ancient hieroglyphics painted on the walls. All were constructed from a combination of clay, bamboo, tarps and rock, and all had sand floors and thatched roofs. Toilets were in small bamboo huts located throughout the grounds, and the showers were set in the center of a poled wooden circle located among the trees –you just followed the path through the poles to get to the center and the showers. Showering was novel and refreshingly delightful.

We shared our space with a variety of little critters who ventured in to keep us company, and after dark, a cacophony of night songs sent us off into dream land.

A large hut, situated at the top of the property and close to the cliff edge, housed the Cliff Café. It overlooked the bay and was a good vantage point to watch the snorkelers in Boston Bay, and the bravest of visitors jump off protruding rocks into the turquoise water. Magnificent frigate birds soared and swirled overhead.

Further along the cliff path, another large, secluded hut (could be the honeymoon suite) housed a kitchen, dining room, and a bathing pool on the edge of the cliff. Only the bedroom was completely under a roof protected from the elements of rain or sunshine.

On both occasions that we visited Great Huts, we spent many hours in the Cliff Café playing cards, reading, and listening to the classical music piped in from the front desk. We inhaled the enchanting fragrances of the flowers, savored the tranquility, and were awed by the beauty of it all.

# GO FOR IT

## Bees, Bees, Bees and Other Winged Creatures

We don't know why but a host of different life species seemed to be attracted to us, or to our apartment – we never figured out which one. Bees were the biggest challenge. One day, our neighbor asked us to make arrangements to destroy a bee hive on the outside of our balcony. We didn't know it was there until he brought it to our attention. His wife was fearful of bees and she complained they were going into their home. Their balcony was only about three feet from ours.

We did a little research and came up with a lady beekeeper by the name of Maureen, whom we fondly referred to thereafter as, "The Bee Lady." She came over on the next Saturday morning to assess the situation and decide on a plan of action. As it turned out, she visited us many times for many hours over the next six weeks before the bee saga was fully resolved.

First, she removed four bee hives from the side wall of our balcony. This in itself was quite a feat, because we were three stories up from ground level! She leaned a long straight ladder up to the balcony and began climbing while Alastair held the ladder for her, and I closed my eyes. She was fearless and quite unprotected when she removed the hives and dropped them into a large garbage bag.

That reduced the number of bees flying around outside, but then the bees appeared inside our apartment! We couldn't figure out where they were coming from. We closed all the windows and doors when we went to work, yet there was a pile of bees, at least a hundred, on the tiled floor in front of the door to the balcony when we came home in the evening. We swept them up, but the next night there was another pile, and their numbers seemed to be increasing daily.

On her next visit, the Bee Lady and her assistant climbed up to the roof, where the water tank was located, and poked around. She found a huge bee hive under the roof's fascia at the back of the apartment. With a smoke pot in hand and assistance from her helper, she removed the fascia boards and six rows of clay tiles to expose the extent of the bee hive and honey cones. This one, she said, was many years old and the cones would have been built by thousands of bees. We looked up at her from a safe distance as she smoked the hive and then, with her bare hands, scooped up handfuls of bees and honey combs and dropped them into large, green garbage bags. She needed six. Satisfied she'd cleaned the area well, she replaced the fascia and roof tiles, and with a huge smile, assured us the problem was fixed. She said she would drop off the bags at the Ministry of Agriculture because the bees might be contaminated with a disease called "Foul Brood." This disease was a major problem in Jamaica, and to control it, bee farmers, whose bees had the disease, had to burn their hives.

But the bee problem wasn't resolved. We still had heaps of bees on the floor by the balcony door every night when we returned home from work, and what was worse, the piles were getting bigger.

And then another strange happening occurred. Honey began dripping from our bedroom ceiling! We put a sandwich sized Tupperware container underneath to collect the drips and within two weeks it was half full.

Maureen couldn't believe it. She shook her head when we showed her the bag of bees we'd swept up the night before, and when we went upstairs and showed her the honey pot, she laughed and laughed, as only a Jamaican lady can.

She went back up the ladder to the roof with her assistant, removed the roof tiles as before and did a further inspection. This time she coated the inside space with a sealing substance. The bees were getting under the roof tiles because of the open spaces and were attracted by the left over honey covering the roof boards. She then brought the ladder into the living room and stuffed wet newspaper into the angled corners of the loft where bees might find a small opening to enter the apartment. Once again she assured us the problem was solved, but it wasn't.

The daily dead bee count continued to rise, and one Saturday, we observed bees flying down the stairs to the main floor. We finally discovered their entryway. It was an empty smoke alarm casing in the ceiling over the stair well. On her fourth visit, Maureen sprayed a chemical into the casing and sealed the unit with plastic and tape. To our delight, this was the last we saw of the bees.

# GO FOR IT

The best part of this experience was in meeting the Bee Lady and the information she gave us about the health benefits of bee products. We researched bee pollen and were so intrigued by what we read that we bought eight bottles of it from her. We took a spoonful of bee pollen every day thereafter until our supply ran out.

Did I mention I got stung by the bees several times during this whole fiasco? Well as it turned out, bee stings were actually healthy for you, if you weren't allergic to them. Bee sting therapy is now gaining recognition globally as a natural medicine to help people suffering from immune deficiencies and arthritis!

*Bees at the door*

*Honey from ceiling drippings*

*The bee lady*

## Other Winged Creatures

We enjoyed watching two doves build a nest and raise a family in the mango tree in front of our condo. The branch they selected for their home extended to within two feet of our balcony, giving us a bird's eye view of their progress. The pair worked diligently to build a nest, and soon after, the female settled down and laid two eggs. One day, two tiny bald headed chicks peeked their heads out from under their mother. They grew quickly into adolescents with fluffy feathers, and then into young adults ready to take flight. It was funny to see them cautiously walking along the branch and looking down as if to say, "You've gotta be kidding – you want me to jump and flap my wings. No way Jose." One did finally, but the other one continued to nervously pace up and down the branch for the balance of the day. Its sibling did fly-byes, seemingly taunting and daring it to follow. It took flight the next day.

The last of the winged creature experiences happened when we traveled out of town for five days. We arrived home after dark, and as was our custom, went to the balcony with a refreshing drink. When we opened the balcony door, a heavy stench of pigeon poop and a huge mess of pigeon droppings greeted us. It was everywhere: on the railings, floor, chairs and table, and there, perched on the top beam of the balcony roof were two pigeons. They glared at us as if we were intruders.

Needless to say we were upset – more so than with the bee episode in fact. It was just so disgusting to look at and smell. We shooed the birds away, closed the door, and went to bed in the knowledge we had an ugly mess to clean up in the morning. It took a good amount of scrubbing with hot soapy water to clean and sterilize the area, but by noon the next day, we'd reclaimed the balcony. To deter them from roosting again, we tied four balloons to the beam. This worked well. We never had another pigeon problem again.

## Treasure Beach

We went to Treasure Beach on the southwest coast, with eight fellow volunteers to get away from Kingston for the Easter long weekend. Not many tourists go to there because it's a six hour drive from Kingston, four hours from Negril, and the last hour or so of the journey is on bad roads. Again, we hired Mr. Mason and his trusty minibus to take us there. We all made our own accommodation arrangements. Alastair and I chose Ital Rest, a Rastafarian retreat, located a couple of kilometers outside town, while the rest of the volunteers opted for rooms in town.

Our journey there was quite the adventure. Since businesses closed at half day before the Easter weekend, we agreed to meet up at noon at the Chez Maria Restaurant to await the arrival of Mr. Mason. We requested a 1:00 p.m. pick-up, but Mr. Mason didn't arrive until 2:00 p.m., and he brought another passenger with him to be dropped off along the way! As it was, we wondered how we'd all fit into the mini-bus and when we'd get there, and now there was one more person and his bags to consider. No one complained to Mr. Mason though. How could we? He was such a sweet and kind man. It was a tight squeeze to load everyone and their luggage into the mini-bus, but we managed.

The new passenger's home was out of town, midway up a mountain on a cliff overlooking Kingston. Although his driveway was unusually steep for a residence, the loaded-down van faithfully made it to the door.

At 3:30 p.m. and now one person lighter, we descended the steep hill, and resumed our journey to Treasure Beach. The next couple of hours went by smoothly. The highway was as good as any in Canada. Some volunteers drifted off into an afternoon nap, others day dreamed, or gazed at the scenery whizzing by.

This all changed in an instant when George noticed smoke pouring from the console between the front two seats! His shouts brought everyone to their senses. Mr. Mason pulled off the highway, and in a panic, we all exited the vehicle, fearing it would burst into flames any second. Mr. Mason wasn't concerned. He inspected the console and concluded the smoke was in fact steam gushing from a broken water hose.

He needed water to refill the radiator so we could make it into Smithville, where he would buy a new hose. Fortunately, we all had water bottles and our contribution of the contents sufficed.

Ten minutes later, we pulled into town. While we explored our new surroundings, Mr. Mason set off on foot to buy a water hose. We thought it would be interesting to visit the coffee factory and coco processing plant, located directly across the road, but they were closed for the Easter weekend. Some of us went into a bakery to buy a loaf of Hardo bread and some cheese.

Laura went off to check out an old, rusted, broken-windowed, tireless, school bus, because she saw a neon "Welcome" sign hanging from the door. We wandered around, and five minutes later, as we neared the old bus, the familiar beat of reggae music and hearty sounds of laughter filled our ears. Inside, Laura had settled on a stool at the bar and was laughing with the owner. Three locals sat at a table playing cards and drinking beer. What an incredible transformation from the outside to the inside, and what an unforgettable sight: a beautiful mosaic tiled floor, six small circular tables with wooden chairs, a finely polished wood bar with four swivel bar stools, and a well-stocked display of liquor bottles behind the counter. What an ingenious way to recycle an old school bus and create a business for yourself out of it!

The patrons were friendly and welcoming, and we settled in as Red Stripe beer and rum punches were quickly served. Conversation flowed easily, and we learned about the growing of coffee and cocoa plants, and the processing factories. One patron, Donolly, was particularly engaging, and told us about his working at the cocoa factory from the time he was twelve years of age until his present age of fifty three. By the time we departed the bus, we had his contact number and a guarantee of a personal tour of both the coffee and cocoa factories, should we ever venture back that way.

At about 6:30 p.m., Mr. Mason returned, fitted the new hose, and advised we were good to go. As we waved goodbye to our new friends, George noticed a sign on the side of the bus that read "We intend to get a liquor license soon."

About three quarters of an hour later, steam hissed from the console between the front seats again. We pulled over to the side of the road and everyone left the van. This time, Mr. Mason was better prepared. He had a back-up hose and three large containers of water. While he completed the repairs, we went to another local bar across the road, and engaged the locals in yet another lengthy conversation about this, that and the other.

We set off and finally around 9:30 p.m. rolled into Treasure Beach. We had difficulty finding the guest houses because the night was black as pitch, there were no street lights, and neither Mr. Mason nor any of us had ever been there. With directions from some locals walking along the street in the village, we eventually found the three guest houses we were looking for.

Ital Rest was proving to be a much tougher quest. It wasn't until Mr. Mason spotted two policemen in a parked car along the dark road on the outskirts of town that we got some good directions. And even then, finding the entrance to the Rasta compound was challenging. We knew there was no electricity, but the blackness of the night made it almost impossible to find. We eventually stumbled across a dirt path from the road amongst the bushes leading to a thatched roofed hut, which was the home of the seventy three year old proprietor. We asked Mr. Mason to stay the night, but he politely refused, and after seeing us safely to our bungalow, set off for the return ride back to Kingston.

Jeannie, our host, expected us to arrive around supper time and had prepared a vegetarian meal of rice and collard greens. We ate it cold, and after exchanging some light conversation, went off to our bungalow with candles in hand.

Ital Rest was a considerable distance from our friends in town. We had choices. We could walk the hot dusty road into town, call a taxi, which would cost about $10 Cdn, or hike along the shoreline taking dips along the way. Which would you chose? As adventurists, it goes without saying, we ventured along the path less traveled. It was delightful – one serene, majestic bay after another. Each little beach was a half-moon shaped cove with pristine white sand and turquoise blue water. And best of all, we were always the only ones there. We swam in each bay to cool our bodies from the blazing heat of the sun, laid out in the sand to dry off, and meditated on the peaceful grandeur. An hour later, we met up with our friends for lunch at Jack Spratts, and stayed the rest of the day to dine and dance to the reggae sounds of a local band.

We liked everything about Treasure Beach: it's friendly, smiling villagers; the beaches; fresh seafood; craft shops and the ever present reggae music that seemed to emanate from every establishment.

On the second evening, Jeanie took us to the home of a local family to listen to twin sixteen year old brothers play the bongo drums. The boys were the oldest of fourteen children in this very poor family, and had never attended school. They worked to help pay the school fees for their younger siblings. Jobs of course were scarce for children, but the boys had a special gift to share with tourists. They'd taught themselves to play the drums, and with such finesse and passion I might add, that I sensed the music was coming from deep within their spirits. They entertained us and about ten other tourists for a good hour with their non-stop drumming. We thanked them for sharing their talents and gifted them a substantial tip.

We visited Treasure Beach three times and every visit was different and special. It was, without a doubt, another one of my favorite places in Jamaica and somewhere I hope to return to one day.

*Ital Rest Cabin*

*Treasure Beach*

*Cove at Treasure Beach*

*Cuso volunteers at the Bus Bar*

## Family Holidays

In August, three of Alastair's grandchildren: Caitlin (17), Andrew (12), and Julia (9) came to visit us for ten days. We first took them on adventures around town, traveling on local buses, route taxis and coaster buses. Each day, we walked out to the main road and stopped to have a chat with George, who lived in an abandoned building on the corner of the street. He was blind as a result of untreated glaucoma, and relied on the kindness and donations of passers-by for food. He in turn cared for his friend who had some mental health issues, as was obvious from his erratic behavior and peculiar manner of dress. He sometimes wore underwear on his head, or put his legs through the sleeves of a t-shirt, and there were times when he sauntered down the middle of the road, causing traffic to swerve to avoid hitting him. George assured us that although his friend behaved strangely and made weird sounds, he was harmless. We trusted his assurance and felt comfortable enough to introduce the children to George and his friend. This pleased him greatly. He said he seldom got the chance to speak with children, and felt blessed with the opportunity to meet our grandkids.

We toured the Bob Marley Museum, wandered through Farmers Markets, and sampled traditional foods at small local eateries. We also took them on day trips to Fort Clarence, Port Royal, Devon House, and Hope Gardens: a sprawling park studded with exotic trees and plants and tropical birds and animals. By mid-day every day, the children were satiated with the new sights and ready to play in the swimming pool.

We took the children on two out of town excursions: one to Treasure Beach and one to Great Huts. Neil, our trusted taxi driver, took us to Treasure Beach, and we went to Great Huts by coaster bus.

Sunset Resort Villa in Treasure Beach proved to be ideal for our needs: we had a self-catering cabin with a fully equipped kitchen, and the resort had a large swimming pool for the children. Most mornings, we swam and snorkeled in the turquoise blue ocean, strolled along the white sandy shore and scrambled over rocks to the next bay. In the afternoons, the children frolicked in the pool or rested in the shade, reading or playing cards. We topped off the day with a walk into town, just ten minutes away, and visited an ice cream shop and Cutthroat Charlie's place.

We'd met Charlie on our first visit to Treasure Beach, and wanted the children to meet him. He was one of those special characters you meet once and remember forever. We introduced the children and left them with him so they could look at his wooden carvings and jewelry. Previously, we'd spoken to the children about how to negotiate a fair price, and they felt they could handle it well. They initially found Charlie a little scary because of his one eye, missing teeth, thick Rasta hair and his thick Patois, but over a number of visits, while they reconsidered what to buy and how much to pay, they warmed up to him. They even went as far as to admit liking him. The name "Cutthroat" reflected the lowest of prices and not a threat of violence. Charlie was most fair to the children in his pricing, and we were grateful to him for treating them so warmly and fairly.

Bill, the owner of the Sunset Villa Resort, was an American, who now lived in Jamaica with his Jamaican wife and two children. One day, he invited us to go "crabbing" with him and his children at night. We eagerly accepted, without knowing what "crabbing" entailed. As it turned out, we weren't prepared for what ensued.

We met Bill, his two children, aged six and nine, and the Resort's night watchman at the pool deck at 7:00 p.m. It was dark, very dark, when we set off on foot down a dirt road to look for crabs, and we only had one wind-up flashlight between all of us. Bill's son wielded a machete, and the night watchman carried a big, burlap bag. We had no idea "crabbing" meant hunting for big blue sand crabs that came out of their holes at night.

We weren't adequately prepared: we didn't have our own flashlight, nor were we wearing closed shoes – we had flip flops. This made us vulnerable in the dark to crab claws that could slice off a toe in one snap. Alastair and I were gravely concerned for the children's safety. We spoke to them about the danger, but they thought it was funny. I thought about ending the adventure right there and then, but I didn't. Everyone was having too much fun. And so we looked on with consternation as the children ran around and laughed. I was tense and waited for a crab incident to occur at any moment, but fortunately none did. The others were fully engaged with capturing crabs by stepping on them (they wore covered shoes), and putting them into the sack. And then my concern shifted to the machete being wielded in the dark by the six year old. Thank goodness Bill finally took the weapon away, after hearing me express concern that we were in more danger from his son than the crabs.

Bill offered to show us how they caught gar fish at night using the machete, but we said we'd had enough excitement for one day. We returned to the hotel for the races. They dumped the crabs from the bags and we picked which one of the ten we thought would travel the fastest and the farthest. The children ran alongside their crab choice, hooting and hollering at them to go faster and straighter. It was loads of fun and we laughed a lot. The children relived their adventure many times over the ensuing years, as they recalled their "night out crabbing."

Another memorable event the children always talk about was the marathon card games of Skip Bo we played under the lights on the pool deck. Though simple in nature, those evenings were magical and so entertaining.

Upon our return to the condo in Kingston, we stayed home for one night to wash clothes and get ready for the next adventure. In the morning, we took the coaster bus to Port Antonio.

The journey was quite the experience for the children. They stared wide eyed at an unfamiliar world while inquisitively watching the behavior and dress of the locals as they struggled, loaded down with bags and packages, to get on and off the bus. They snapped pictures of unfamiliar sights, and gazed in awe at the majestic beauty of the Blue Mountains, while apprehensively keeping an eye on how close the bus was to the cliff edge and the deep valleys below. Although the packed bus was uncomfortably hot, humid and noisy, the children didn't seem to my mind. They didn't complain, moreover, they seemed to thrive on the novelty and energy of it all. After the long and arduous bus ride, we were much in need of tranquility and the freshness of sea, sun and shade, and we had the perfect place in mind.

Frenchman's Cove was a stunningly beautiful Jamaican treasure, and only a half hour route-taxi ride east of Port Antonio, and on our way to Great Huts in Boston Bay, our final destination. The children were in their element with few people on the beach, clean, warm, white sand, and gently lapping azure blue waves. We settled down under the shade of a tree, and for the next hour or so, luxuriated in the cove's splendor.

Now relaxed and eager to get to Great Huts, we took the path to the main road and began walking towards Boston Bay. It was hot. The sun was bright, the air heavy, and we were getting wearier and sweatier by the second. The children were tired and hungry and moving slowly. We didn't have any liquids left to drink! We soon realized we'd misjudged the distance: it was much too far to walk to Boston Bay and so we needed to get a ride, and soon. We didn't have a cell phone, and after watching one route-taxi after another zoom by, we realized there was a problem. Taxis didn't leave Port Antonio until they were full! What to do? There was no one on the road to ask, no hotels or houses in sight, and we had no other ideas on how to resolve our dilemma.

Just as we were beginning to stress about the situation, a vehicle stopped. It didn't look hopeful because it was a small taxi already full with four passengers: one in the front and three in the back. Yet, the lady taxi driver offered to take us to Boston Bay. She'd guessed our dilemma and the reason for it, and took pity on us. We must have been a pathetic sight – two obviously exhausted and stressed elders with three overheated children at the side of a lonely highway.

One very accommodating lady climbed into the trunk to enable us to fit into the car! What does that say about Jamaican hospitality? I can't see that happening in Canada. The other three passengers squeezed together to free up a seat for our oldest grand-daughter, and the other two children sat on knees. Yes it was tight – six people crammed into the back seat, two passengers in the front bucket seat and one good-hearted soul in the trunk, plus the driver and our back-packs. In retrospect, we weren't surprised with this development because it was another example of the kindness and generosity we frequently experienced from the Jamaican people.

At Great Huts, our accommodation was a large tent with three beds, complete with mosquito nets to keep out the geckos, frogs, bats and bugs. We didn't want them climbing, hopping and sharing our space when we slept. Great Huts intrigued the children – they'd never experienced anything like it. To them, it was one huge playground. They chased each other and played hide and seek as they explored all the nooks and crannies of the place. They lounged in the unique clay, vine and bamboo chairs, and swung in hammocks located throughout the property. It didn't take long for them to muster up the courage to jump off the cliff into the water, and then there was no stopping them – that's all they wanted to do, over and over again. And when they did this, Alastair and I went into the water near where they landed in case there was a problem. But there were no mishaps.

The laid back environment contributed to our relaxation and enjoyment. Kitchen staff only appeared at meal times, and except for the odd gardener, we seldom saw any other staff. There were no sales desks selling this and that day-trip package, nor any organized daily activities with blaring music you get at more upscale resorts. All guests went about their business quietly. Even the snorkeling in Boston Bay, which incidentally was some of the best snorkeling we've ever experienced, was free. You just helped yourself to the snorkel tubes and masks sitting in buckets at the top of the steps down to the water when you wanted to go snorkeling. Meals were included in the price, and though simply prepared, were delicious and ample.

The children fueled up on the cooked breakfasts of eggs and toast every morning, and usually found something on the supper menu to eat. They showed much independence and bravery in tasting new foods, and even tried the jerk chicken, once, but it was too spicy hot for them. And I might add, too hot for many guests from other parts of the world I spoke to. But it was the fresh fruit that was always available that I most remember about Great Huts – the oranges, mangoes and pineapples, and their fruit smoothies were to die for. And of course, the Blue Mountain coffee with breakfast and Red Stripe beer with supper, were always pleasing meal supplements.

We played Skip Bo immediately after supper every night and laughed until our tummies hurt. Isn't it amazing how fiercely competitive children can get with the right setting and mood? It became an evening ritual with them – there was never any question as to what we'd do after supper. At 8:00 p.m., a local cultural group took the stage and entertained us with singing and dancing. The children usually joined in, banging bongos and dancing, and we had a lot of laughs. Our time at Great Huts with the children was another of my favorite memories of our time in Jamaica.

At the end of ten days, with tears in our eyes and sadness in our hearts, we took the children to the airport and waved goodbye. It was a privilege to have had them, and we were proud to be the grandparents of such adventurous, well-mannered and fun young people.

## Best Christmas Ever

In October, we learned that my children were coming to visit. My two daughters, Gabbi and Shannon, Gabbi's husband, Ken, and Shannon's newly adopted daughter, Ashlyn, were coming to spend two weeks with us over the Christmas holidays. How wonderful and serendipitous was that! Being posted overseas provides an unusual opportunity for friends and family – another great bonus to volunteering.

As the adoption had taken place while we were in Jamaica, this was the first time we'd met Ashlyn, who was eleven months old. It was to be an exciting, different type of Christmas. There would be no presents, tree or snow, but I would get the most wonderful gift I could ever wish for – their company in Jamaica. This was truly to be the best Christmas ever. Although I'd kept in close contact with Gabbi and Shannon via Skype, I hadn't fully realized how much I'd missed them until they walked through the airport doors and I put my arms around them.

Ashlyn's adoption, only four weeks prior to their arrival in Jamaica, fulfilled Shannon's greatest dream to become a mom, and she was blossoming in her new role. I was so excited and happy for her. The twinkle in her eyes and smile on her face when she looked at her daughter brought back the profound sense of love I felt so many years ago when I held Shannon as a new born babe in my arms. The feeling was incredible. I silently thanked the Universe for bringing this little girl into my daughter's life, and into mine too.

Ashlyn was going to be loved by so many in her new large extended family. I gazed upon her silhouette as she clung to her Mom, and my eyes filled with tears of happiness. Her beautiful blond hair, blue eyes and petite frame were precious. My heart felt as if it was going to burst with the love I was feeling for Shannon and her miracle baby. A line in a song came to mind: "Never in the world can there be too much love."

I also had the good fortune to become a surrogate Grandmother a number of years earlier to two little girls: Alyssa aged three, and her sister Shawna, aged one. They were my great nieces and didn't have a grandmother. One day, Alyssa clasped my face in her little hands and asked, "Are you my Grandma." I gladly replied, "Yes I am." And since then, my bonding and relationship with them has strengthened. Although Alyssa and Shawna live three hours away, we keep in touch and visit when we can. They will forever be my Granddaughters in my heart.

Ashlyn was a treasure: such a sweet, joyful, little baby who warmed easily to people, yet made sure her new mom wasn't too far away. Her need to keep mom within sight was understandable as they were still in the early stages of bonding. As much as I wanted to hug her, kiss those beautiful little cheeks, and nuzzle into that unforgettable baby fragrance in the crook of her neck, I restrained myself until she showed signs that she wanted to be held by me. It was a lot of adjustment for her – new faces, new sleeping arrangement and feeding schedule, and a different climate, but overall she was taking it in her stride and adjusting well.

Napping was a little tenuous at first, but once we discovered she liked reggae music (she quickly fell asleep listening to a loud version of "One Love" when we visited the Bob Marley Museum), she didn't have a problem after that because there was reggae music everywhere we went, day and night.

We were so familiar with Kingston by now and in acting as tourist guides that putting together an itinerary for my children to see the best of Kingston was easy. We spent the first week visiting Hope Gardens, Devon House, the Farmers Market, Fort Clarence and Port Royal and the Bob Marley Museum. For the second week, we decided to rent a car and drive around the island, stopping at some of our favorite places on route.

This was exciting for Alastair and me too because we'd often spoke about renting a car to see parts we hadn't yet seen. However, driving on the left side of the road on those unpredictable motorways was challenging. One time in particular comes to mind when we followed the directions of a local person to take a short cut. It was raining heavily, and the road we were instructed to take gradually morphed into a tight gravel path. And then it went up a very steep hill. It was a sheer drop down the hillside on the passenger's side. We all gasped (and I closed my eyes) as Alastair steered the car, which slipped several times on the wet gravel, ever so carefully up and over the top. We laughed about it later, but at the time I know I was scared silly. I imagined our year in Jamaica ending in a car accident at the bottom of the hill.

We took them to Treasure Beach. This was our third visit and the second time we stayed at the Sunset Villa Resort. We got a larger cabin with three bedrooms this time. Bill had decorated the place with festive Christmas lights and ornaments, and cooked a full Christmas turkey dinner with all the trimmings for hotel guests, family and friends. It was a most memorable Christmas dinner.

We'd talked a lot with Gabbi, Ken and Shannon about how great the beach was at Treasure Beach and wanted them to enjoy it as we had. Imagine our surprise when we went to see the sand and found only a rocky shore instead. We couldn't believe our eyes, in fact, we didn't even recognize the beach. Treasure Beach was located in St. Elizabeth's Parish, and sadly, the area had been hit by the tail end of hurricane a month earlier. The high winds had picked up the sand and blown it inland, and the huge waves had carried the rest back out to sea. The shoreline now consisted of bare rocks. Waves rushed in and filtered back through the crevices to the sea. Nature's power to so quickly change shoreline contours was mind-blowing. Bill said that it had happened quite a few times in the past, but subsequent storms had brought back the sand, and they would again, one day. We eventually found a good beach further down1 the shore, but had to climb up and over rocks for a distance to get to it.

We couldn't come to Treasure beach and not see Cutthroat Charlie. He was pleased we'd remembered him, and he remembered us and asked about the grandchildren. Gabbi, Ken and Shannon liked his work and bought a number of carvings as gifts. I was surprised to see him the next day on the hotel grounds looking for Ashlyn. He'd made a special beaded bracelet for her. This simple gesture of a good hearted man filled my heart with joy.

We continued on our way west and north, enjoying the beauty of the hills and countryside. I'd pre-booked a guest house in Negril, but we couldn't find it, despite asking many local people for directions.

It always surprised me how the Universe appeared to send us what we need, when we need it. As we drove out of Negril and up the highway a little way, we saw a sign that led us to a very special place called Half Moon Beach, Green Island in Hanover. This private location had it all: great beach, comfortable small cabins, and a little restaurant serving lobster at a very reasonable price. We each had our own cabin overlooking the ocean and the swell of the waves rolling into shore lulled us to sleep.

We traveled to Montego Bay the following day, but chose not to stay there because it was too busy and noisy. There were so few public beaches because the "All Inclusive Hotels" owned most of them. So we continued our drive across the north coast to Ocho Rios and stayed there for two nights in a private guest house just outside of town. As this town was also a cruise ship port, it had large "All Inclusive Hotels," many pricey tourist attractions, such as "Dunns River Falls" and the "Swimming with Dolphins," exhibit, as well as the usual plethora of shops selling brand-name products. We considered going on a hike to swim in a cascading waterfall, but unfortunately it was raining hard. We decided to return home to Kingston through the Blue Mountains.

It was an interesting ride and a little scary at times. Driving on the left side of the road with the driver in the right front seat was disorienting. There seemed to be so little room between the car and the edge of the cliffs, or the vehicles traveling in the opposite direction, especially the big buses and trucks. We were all relieved, especially the driver, Alastair, when we finally pulled into the compound at Marley Manor, but no one said anything.

The children admitted they enjoyed the organic and homelike atmosphere of staying at guest houses instead of hotels and that pleased me. Maybe they'll consider this type of holiday in the future.

They returned home on December 29th. I was sad to see them go, but I would see them again in three weeks when we returned to Canada. My daughters are grown women living independent lives, but in my heart, they will forever be "my little girls." And with every passing day, I'm proud to be their mom.

## Forever Friends

When you volunteer and leave your family and friends, you develop a surrogate circle of support with your fellow volunteers. As you share your life with them, you bond and develop new friendships. They become the people you turn to in times of need: a shoulder to lean on when overwhelmed with homesickness; a comrade to brainstorm ideas with; a non-judgmental listener when frustrations rise to the top; a nurse to help you through a sickness; and so much more, but most of all, they are the people you play with.

Alastair and I spent many hours enjoying their company, some of whom are now life-long friends. We enjoyed meals at one another's apartments, shopped at local markets, attended special events, and celebrated birthdays and holidays together. Robert and Dominic came to our house once a week for two months to give us private Spanish lessons. As well, we spent numerous weekends together on out of town adventures.

Towards the end of our placement in Jamaica, one such trip stands out in my mind because it made me realize how much I appreciated the uniqueness and diversity of our volunteer family. George and his wife Bunty, Dominic, Dom, Robert, Delphine, Alastair and I went to Holywell National Park in the Blue Mountains for a weekend of hiking and to savor the beautiful sights in the Jamaican countryside.

We rented a two bedroom cabin. The group generously agreed to let George and Bunty and Alastair and I have the bedrooms while they slept on sleeping bags on the floor in the living room. A brick fireplace separated the kitchen from the sitting room. Outside on the porch, was a large stack of firewood for our use.

We lit a fire the first night because it was chilly and there was insufficient bedding for eight people – none of us had given any thought to how cold it might get because at sea level, where we lived, it was hot and being cold didn't enter our minds.

Sometime in the middle of the night, we suddenly woke up and heard Dominic screaming "Get out. Get out." The fireplace had malfunctioned and the room was thick with smoke. It had caused Dominic, who was sleeping closest to the fireplace, to wake up and take action. Without fear for his own safety, he grabbed the burning logs, ran with them to the door, and heaved them onto the lawn. He shouted for us to get out, to open all the windows, and to go outside for fresh air. This obviously was an extremely dangerous situation because of the danger of carbon monoxide poisoning, and we wondered what might have happened if Dominic hadn't woke up when he did and responded so quickly.

On the second evening, we sat outside on the veranda and gazed down on the city lights of Kingston, far, far below. Occasionally, distant sounds of reggae music drifted up, adding to the Jamaican ambiance. We chatted about our time in Jamaica and how the volunteer experiences had impacted our lives. As I listened to their stories, my heart filled with a warm tenderness of gratitude for having these people in my life. It was evident that they too appreciated the opportunity to contribute their skills to improve the lives of others. I had much respect, not only for this group of dynamic people, but for all volunteers who give so generously of themselves to someone or something outside their own lives. This event re-enforced my belief that all people were inherently good, and have compassion that they demonstrate in their own ways and means in their own time.

On the last morning, five of us ventured out to hike a trail into the forest. Alastair stayed behind because his ankle was troubling him, and he didn't want to aggravate it by walking.

Upon our return, he described a profound experience he had while we were away. He told of how he sat on a bench in the garden overlooking the valley to meditate. In the middle of his internal trance, came a pleasing, haunting melody piped with a flute. The music surrounded him and wove sweetly into his breath, swaying him ever so gently like a child in a cradle. When he opened his eyes, he saw Dominic absorbed in a musical interlude with nature on his flute. He was ad-libbing: playing notes and melodies as the mood took him. Little did we know that this quiet, unassuming man had such a powerful talent, which he chose to share with Alastair in that special place and time? Dominic has since given us a CD of his music, recorded in cathedrals and other unique places with great acoustics that he discovered on his vast travels in the world. Alastair will be forever grateful for that precious memory of a time spent meditating with Dominic's gift of music.

## Homeward Bound

As we prepared for our departure in mid-January, I reflected upon the past year and what I'd learned. For sure, it had been the most inspirational, exciting and rewarding year of my life. I was meeting all the needs identified in my Personal Statement of Fulfillment, and by sharing my skills and abilities, I was truly helping grass root NGOs and their staff to build up their capacities and effectiveness.

What was most meaningful, and where I felt I had the most impact, was in my working with co-workers to help them strengthen their abilities and confidence. At the same time, I was also learning how to better communicate and relate to people of other cultures, and my capacity to understand, to be tolerant and patient had grown considerably. I could now readily accept "what was," and I felt loving kindness and compassion for others less fortunate than I, and for their life problems. I was living in the "now," participating fully and intentionally with greater presence and passion than ever before. I was totally engaged with a purpose far greater than anything I'd done in my paid work life. I was living my dream.

It had been such an amazing learning experience for us, and although we were preparing to end this placement, we were already mulling over when and where to go again, with whom and under what circumstances.

But for now, we were enjoying each day. We valued the opportunity to have met so many wonderful people, and the chance to explore life in a tropical paradise, while increasing our awareness, understanding and appreciation of life.

Our placements finished in mid-January and we returned home. Back in Canada, we each lived independently with our children because we were "homeless" people. While considering options on how to continue our new adventurous lifestyle, we caught up with the lives of family and friends, and used the opportunity to spend as much time as we could with the grandchildren.

As to options, one that interested me greatly was in teaching English as a second language in another country. As Alastair had already completed the TESOL course and had his certificates, I decided to enroll in this program and get mine. With TESOL, in addition to the basic course, you get to select the target audience you'd like to teach. I chose "children." Alastair chose "business people." I successful completed the course, and began day-dreaming about where we might go in the world to teach English.

But before we did any serious research on teaching positions overseas, wanderlust took hold and we decided, quite impulsively I might add, to budget back-pack through Central America for four months starting in March.

## Guyana
### (by Candas)

When we were in our third month of backpacking – we were in northern Nicaragua at the time – we decided to reapply with Cuso. Imagine how thrilled we were when we looked at their web site and immediately saw two positions in Georgetown, Guyana that seemed to be a great fit for us. We immediately applied, and whenever we could get internet service, checked up on the progress of our applications. The matching process took about six weeks with interviews and paperwork. It was manna from heaven. The Universe was looking after us once again.

We researched Guyana on the web. It was a lesser developed country than Jamaica, but with a similar history of political turmoil and violence. This didn't deter us from wanting to go there. Guyana means "Land of many waters" in Amerindian. The country had vast rain forests dissected by numerous rivers, the main ones being the Essequibo, Demerara and Berbice, and many waterfalls, the most notable of which was Kaieteur Falls on the Potaro River.

There were five distinct areas in Guyana: the lush swampy plains along the Atlantic coast where 90% of the country's inhabitants live; a dusty white sand belt, rich in gold and diamond deposits; dense rain forests located in the interior; verdant flat savannah grasslands along the Rupunini river in the south, and the Interior high- lands comprising the highest mountains in the country. Mount Roraima on the Brazil-Guyana-Venezuela border was the highest.

## Georgetown, Guyana

I was accepted for the position as the National Disability Volunteer Coordinator (NDVC) for the National Commission on Disability (NCD) in Georgetown. Alastair was offered the position of Organizational Management Advisor for the Young Leaders of Agricola (YLA).

Our placements started in August, less than two months after returning from our back-packing adventure. There was a lot to do: pre-departure documents to fill out, medical examinations to be undertaken, vaccinations to be had etc. We spent as much quality time as possible with our families, in the knowledge we'd not see them for a year. We doubted if any of our family members or friends would visit us in Guyana, as they did when we were in Jamaica, because it was so remote and so relatively far from Canada.

I was grateful for yet another opportunity to do the work I most enjoyed: to contribute to something I felt worthwhile; to be engaged in a new community of people; and to continue learning about life.

## Guyana

Guyana was first colonized by the Netherlands, and then by the British. It became a British colony, known as British Guiana, and remained so for over 200 years until achieving its independence in 1966. It's the only country in South American where English is the official language. As was the case in Jamaica, the locals spoke English with a strong Patois that was sometimes difficult for us to understand. The Patois in Jamaica was a type of creole – a blend of English and African languages – whereas in Guyana, the Patois had also been strongly influenced by the Dutch language, and that made it incomprehensible to us.

Guyana was bordered by Suriname to the south, Brazil to the southwest, Venezuela to the west and the Atlantic Ocean to the north. It had one of the largest unspoiled rain forests in South America, parts of which were inaccessible by humans. It also had a huge diversity of wildlife, including undiscovered species, as well as rarities, such as the giant otter and harpy eagle. Despite Guyana's untouched tropical beauty and recent investment in Eco-Tourism, it was not a destination considered by most tourists.

## In-Country Training

We arrived in Georgetown, Guyana on August 12th. Little did we know that on the same plane were two other volunteering Canadian couples, Ian and Maryisa, and Anne and Andy, who would eventually become life-long friends?

During the first two weeks, we received our In-Country training. It comprehensively covered every aspect of Guyana we needed to know – its history, culture, food, and the work Cuso had done over the past thirty years. We also learned about the programs and organizations we'd be working for, and an overview of the work we'd support. Upon completion of the intensive training, we felt ready and excited to get started.

Throughout the training, we stayed at a comfortable, though-crowded, guest house called Rima, with fourteen other volunteers —seven Canadians, five Brits, a Dutch lady, and a young man from the U.S.A. Three were couples around our age (two from Ontario and one from Britain). The American and one of the Canadian volunteers were young men in their early twenties who were there as part of a Co-Op program for the International Development Degree at the University of Toronto.

The Rima provided a comfortable, clean room and three scrumptious meals a day. It was a twenty-minute walk from the two hundred and eighty mile Sea Wall running along that part of the Guyanan coast. Unfortunately, Georgetown is located where a number of huge rivers converge and the river sludge, deposited at the river mouths, turns the normally azure blue Caribbean Sea into a brown muddy color. No one swam in it.

## The City and Its Environs

Georgetown is an old colonial town. The original buildings were made of wood, and though many had been beautifully restored, the majority was in varying states of decay. Newer buildings were taller, but only about three or four stories high. Some main streets had tree lined walk-ways down the center, separating the traffic moving in either direction. These tree lined boulevards reflected a time gone by when Georgetown was called "The Garden City." Unfortunately, the landscaped, beautiful gardens had long ago lost their luster, and were now in dire need of revitalization.

The city had been designed by British engineers with a system of culverts and canals running adjacent to the roads to manage the flow of ground water. Because of its tropical location, Guyana received torrential rains and hurricanes and flooding was a regular occurrence. The unfortunate part was the culverts were now filled with stagnant water because they were clogged with plastic bottles, Styrofoam and plastic bags. When it rained, it poured cats and dogs, and because Georgetown was at sea level, the water in the blocked culverts and canals rose up over the roads and flooded the streets, within a matter of hours.

This was a major concern when it came to mosquitoes and health related issues, but managing garbage and recycling plastics didn't appear to be a priority for the Guyanese government, for whatever reason. We therefore had to sleep under a mosquito net, wash the net regularly with a substance to deter mosquitoes, spray repellent in the rooms before leaving for work every day, as well as dowsing ourselves with bug spray day and night. Yuk – all those chemicals! Well, better safe than sorry. Malaria wasn't a concern in Georgetown, but Dengue Fever was, and that was nasty, as Alastair could attest to after his Bangladeshi experience. Cuso provided us with the net, sprays (eight bottles a month), fire alarm, first aid kit, personal alarm and a flashlight. They also provided an allowance to buy a bicycle and lock if we so desired, but we decided not to cycle because the drivers of buses and cars were too scary. They didn't seem to have any consideration for cyclists, or pedestrians for that matter. Bike accidents were common.

The city water wasn't drinkable so one of the main manufacturing industries in Georgetown was the production of filtered bottled water – thus the discarded plastic bottles in the culverts. Cuso also provided us with a water filtration system to generate drinking water, and instructed us to wash all vegetables and fruits in the filtered water. City water also contained a lot of iron requiring the use of bleach to purify the water, wash clothes and for general household cleaning.

Guyana's population is less than a million (about 760,000), 90% of whom live on the narrow coastal plain. The population density is more than a hundred and fifteen inhabitants per square kilometer. Compare this to the population density for the whole country at less than four people per square kilometer and you get an idea of how remote the interior is. That's where most of the Amerindians live.

Three main ethnic groups comprise most of the population: East Indians – about 43%, Africans – about 30%, and native Amerindians – about 9%. Multi-racial groups represent about 17% and Europeans and Chinese account for less than 1%. Africans were first brought to Guyana as slaves to build the railroads and manage the plantations, and when slavery was abolished, East Indians and Chinese were brought in as indentured laborers to replace them. Many people had a blend of English, Dutch, French, Spanish, African, Chinese or East Indian in their parentage, and babies from this group were referred to as "cook-up:" a culinary metaphor reflecting the mixture.

80% of Guyana was undeveloped rain forest. The interior only had one major road running in a north east to south westerly direction from Georgetown to Lethem on the Brazil border. It was largely unpaved and quite frequently not travelable because of flooding. To get to many Amerindian villages or small towns in the interior one must travel on a river.

Some villages were developed to support miners, known as "pork knockers," working in the gold and diamond mines. The origin of the term "pork knocker" wasn't entirely clear, however, some said it referred to the miners who came back into town after a long shift in the mines, and knocked on people's houses, asking for pork! They must have had ravishing appetites!

Our Journey

## Staying Out of Harm's Way

As in Kingston, security was a big issue in Georgetown, and we received a lot of information about how to keep ourselves and our property safe – "don't resist, just hand it over." We were advised not to go out at night after 7:00 p.m., and if there was an occasion when we had to, then we were to take a taxi, even if it was only a few streets away. Locals and volunteers told us there was a good chance we'd experience a mugging while in Georgetown! The research we did prior to departure didn't indicate this level of mugging frequency. The robber's intention, they said, wasn't to hurt anyone, but only to take their "stuff" (money, cell phones, IPods, computers and even cigarettes). Somewhat a relief, I guess, but still. Hmmm, with all that foreboding news, we wondered if we'd made the right choice in coming to Guyana. The poverty of the people was obvious, with many beggars on the streets, especially persons with severe disabilities. It was so unlike our experience in Kingston, Jamaica.

The brutal reality of the security issue hit home the second morning of our orientation. Two volunteers arose at 6:00 a.m. and went for a walk along the sea wall. They hadn't been gone ten minutes when a man greeted them politely, and then advised he had a gun in his pocket, so they were to give him everything. They told him they didn't have anything. He patted them down and found a pack of cigarettes and a lighter, which he happily pocketed. The smoker protested and complained he wasn't a very nice person to be leaving her without a cigarette so early in the morning. The robber apologized, took one out of the pack, lit it and gave it to her, with his best wishes for a pleasant day. He then merrily headed off with a swagger and a song. The poverty level was high with few employment opportunities, resulting in people desperate to feed their families. But aside from the security issue, the Guyanese people were friendly, helpful and happy. They loved to laugh and enjoyed meeting people from other cultures.

## Moving In

Unlike our placement in Jamaica, we didn't have to find our own accommodation. As flats were apparently hard to come by in Georgetown, Cuso took on this responsibility and assigned apartments to volunteers. Our quarters were located down an alley behind a house on the front of a street. There were three apartments in our building. We moved into the second floor – the first and third floors were occupied by other volunteers (both married couples). A seven foot concrete wall and a locked iron gate protected the house. Inside the walled enclosure, there was no grass or trees, no inviting terrace or balcony, such as we had in Jamaica – it was all concrete. The houses on either side were only about ten feet away. The humidity and moisture in the air tainted the concrete with what looked like black mold. Some owners painted the concrete, but the upkeep of re-painting every year was such a chore many home owners didn't bother, and left the walls in their natural grey concrete state.

Unlike Marley Manor, this location was always noisy. The houses were close together, and we left our windows open most of the time to air out the rooms and let in fresh air. The noise from our neighbor's comings and goings, including their never-ending music, drifted into every room, at all hours of the day and night. In a one block area, every 3rd or 4th dwelling housed a small convenience store selling beer, and offered a couple of tables at which to sit, drink and listen to music. These shops stayed open late and the music increased in volume with each passing hour.

The apartment was almost bare of furniture. We had a stiff, uncomfortable, three-seated fake leather couch, a kitchen table and four chairs, a bed in each of the two bedrooms and a bookcase. That was it for furniture. There were no seats for guests, no coffee table for cups or plates, no television etc., but we did fortunately have internet that we shared with the other two tenants. This bare bones situation would have sufficed if we were in our early twenties and went out a lot, but that wasn't us. We were in our sixties, and as we intended to stay home a lot because of security issues, we felt the need to create a more home-like environment.

We checked with Cuso to see if they had a storehouse of furniture left by departing volunteers, but they didn't. In Guyana, volunteers either gave their household items to incoming volunteers or to Guyanese friends and co-workers. We asked our landlord if he would provide some of the items on our list, such as a coffee maker, toaster, kettle and pillows. He said no: he couldn't afford to. In discussing the situation with other newly-arrived volunteers, we were surprised to learn that some had brought a second suitcase filled with linen and household items, because they knew the furnishings would be minimal.

It was only then that we realized how good we'd had it in Jamaica – we'd been spoiled. It was a one-off. Most volunteering placements were not like that, Alastair said, reflecting upon his Bangladeshi days. It was going to be a much more austere experience than Jamaica that was for sure. No problem. The next Saturday we went to the shops and bought what we needed to make our place more like home because, quite frankly, we could afford to so why not.

Because there was no washer or dryer in the building, we had to hand-wash everything in the kitchen sink, take it downstairs, and hang it on the clothesline in the yard. We did this for the first few weeks, but at our age, did we want to spend our weekends on this mundane, manual activity? We were more interested in getting out and about in daylight hours, and so found a neighbor who gladly took in our washing for a small fee.

Weekly grocery shopping was a most pleasant event, albeit hectic and time consuming. We first went to a local Farmer's Market – there were two to choose from – for our fruits and vegetables, and then walked over to the grocery store to get the rest of our provisions, and took a taxi home.

The Farmer's Markets were sprawling, noisy, smelly and colorful affairs. We soon learned they were also highly competitive, and this changed our routine and helped us spend less. We spent the first hour wandering around stalls, checking prices and quality. And then, with minds made up, revisited the vendors and made our purchases. As with most people, we took great satisfaction in getting something at the least cost, and felt we'd been had, if we found a cheaper price somewhere else later.

Once we were familiar with the layout and the vendors, and had exchanged pleasantries with certain stall owners, we developed a rapport and an allegiance with the ones we liked the best, even though they weren't always the cheapest. Knowing their circumstances no doubt influenced our bias, but that was life, and we were okay with that. We looked forward to seeing their smiling faces on a Saturday, and if we paid a little more than the competition, we were okay with that too.

It's strange how one adapts to living in a country with a different economy. You unknowingly change your mind set and function accordingly. In Canada, we'd not give a second thought to stopping at a coffee shop mid-afternoon to order a coffee and a donut for three dollars. But three dollars is a pile of money in most undeveloped countries, where most people only earn two dollars a day! So, when we were shopping around the market, looking for a cheaper price, we're talking pennies. Would anyone in Canada spend a half hour in a congested, hot market to find a mango two cents cheaper? Not on your life, but volunteers do when they go to another country. We raised this point several times with other volunteers and they smiled. They admitted they didn't look at their spending this way, but it was good food for thought.

Beer and rum were plentiful, but there was a limited supply of wine and ground coffee. That was surprising since we were surrounded by countries producing fantastic varieties of both. A bottle of imported wine averaged about $20 Cdn. At that price, we didn't often treat ourselves.

It took about twenty five minutes to walk to work and ten if I took a mini-bus. Though I loved walking, I didn't like to in Georgetown because it was too dangerous. There were many bus, cycle and pedestrian accidents in the city every day. I saw several instances where mini-buses had crashed into culverts or buildings, and heard many stories of pedestrian fatalities.

## Getting Down to Work

I was excited and enthused about my placement because I felt it was a great match with my forty years of experience in the social services sector. After orientating myself to the organization, I adjusted my personal work plan to reflect what I believed could be accomplished.

My main objective was to work alongside my in-country counter-part to develop and implement a National Disability Volunteer Program (NDVP). Guyana had passed a National Disability Act in 2010, which gave all persons with disabilities (PWD) the same rights as everyone else, but the general public didn't know this. The NDVP's goal, via trained volunteer facilitators, was to make the country aware of the Act and to explain the work of the National Commission on Disability (NCD). Another objective was to register every PWD in Guyana on the NCD's National Registry so they could get Public Assistance, as well as access to other health and rehabilitative services. There were an estimated 50,000 PWDs in Guyana.

The three volunteers I worked with were PWDs: Ganesh was blind; Simone was a wheelchair user; and Rennata had short arms. Each had extensive experience in the disability movement and was actively involved in different disabled person's community groups. They were skillful and confident advocates, and impressed me with their facilitation skills. I knew from the beginning they'd do a great job in sensitizing the public.

They volunteered three days a week and received a stipend, equivalent to about $4.50 in Canadian dollars for three day's work. What would that get you in Canada? A coffee and two donuts! They also received a transportation allowance to cover their travel costs to and from the NCD office. Most had no option but to take taxis because there were few sidewalks, few traffic lights and many pitfalls, such as ditches filled with stagnant water and garbage. There was no infrastructure in place to assist PWDs to access buildings, or to move freely around Georgetown.

## The "Just Now" Challenge

As was the case in Jamaica, I faced many work challenges, the most severe of which was the lack of resources. At various times there was no ink for the printer, no paper for the copier, no water in the taps to flush the toilets, no money to replace the broken toilet seats in the handicapped and women's washrooms, no hydro to keep the fans going and keep us cool, and oftentimes, no water in the cooler to drink. Let's just say a lot of bonding took place as we waited for one thing or another to happen before resuming our work.

There were never enough resources, and certainly one had to adjust to the "just now" approach to get things done. "Just now" could mean in a little while, tomorrow, next week, or never. One learned when and how to ask again, or to remind people. One never assumed "just now" would happen any time soon.

The same "just now" approach was all too evident within the Government who'd passed the Disability Act. Other than setting up the NCD, the government hadn't provided funds to implement the Act. The process was slow and arduous. In Guyana, most people didn't know PWDs now had the same rights as them, and that it was illegal to deny them services and care. For so long, PWDs had been denied, shunned, discriminated against, and even locked away behind closed doors because families were ashamed of them.

I felt privileged to support such a vitally important project advocating for such basic rights as: the right to education; to health services; to vote; to play sports; and to enter buildings and receive services just as any able bodied person could.

## Alastair's Work

Alastair was enjoying his placement as well. He worked with ten amazing young people, ranging in age from sixteen to mid-twenties, who lived in a scary place called Agricola, a small Afro community on the outskirts of Georgetown. It had a horrendous history of gang violence, resulting in a stigmatized reputation for anyone from there. Young people in particular had a difficult time in the city when they applied for a job, because city folk feared anyone from Agricola. This was what the youth wanted to change. They called their organization, "The Young Leaders of Agricola" (YLA).

Alastair's primary job objective was to help them develop a five year plan to heal their community and eliminate the stigmatization to benefit future generations. Another objective was to assess the long term viability of YLA as a partner with Cuso.

He had many challenges, but the main one was security. Cuso insisted he should never travel to Agricola by himself! It was a risk they didn't want him to take. Some taxi drivers even refused to go there for fear of being robbed! The first few times he went into the community, Milton, the group leader, took him around and introduced him to as many residents as he could find, as well as to shop keepers, and even the drivers who drove the buses into Agricola. Milton told the people it was their responsibility to take care of Alastair and to ensure no harm came to him. He was also taken to meet a couple of the "bad boys" – those who'd caused problems in the past – so they knew who he was and what he was doing there. They too had to ensure his safety. After his meetings with the locals, Alastair felt quite comfortable and safe. He tried to always take a certain bus, driven by "Shoeface," who would save a seat for him, even when the bus was full. The community greeted him warmly.

His work was also frustrating. He could only meet the group on certain evenings in the week and on Sunday afternoons after church, because most members had jobs, worked a regular Monday to Friday schedule, or were full time High school students. The passion of these youth, who were so strongly committed to changing their community for the better, impressed Alastair. He agreed to make himself available to them on their time schedule, even if it meant traveling to and from the community in the evening. Cuso didn't like this, but Alastair didn't have another option and was okay with it.

## Pegasus Reunited

Once again the Greek horse looked down fondly on us and smiled. We were reunited with our old friend, Pegasus, and as in Jamaica, this was a hotel with a pool, a lobby television to catch up with the outside world, and a welcoming ambiance in which to connect with other volunteers. But unlike Jamaica, we had to pay for this privilege by joining their Athletic Club. The price, however, was reasonable, and an expense we were willing to pay in exchange for the sense of well-being it provided.

Alastair and I met up after work around 5:00 p.m. for a refreshing swim two or three times a week. The Pegasus became the focal point for our weekend, and all activities revolved around it. On Saturdays, after grocery shopping, we'd head over there to meet up with fellow volunteers and lounge poolside, enjoying the water and the surrounding gardens. Once in a while, we'd treat ourselves to the extravagant buffet, laid out at poolside, and pretend we were regular tourists at the hotel. On Sundays, we left home early, as we did in Jamaica, to have the entire pool and facilities to ourselves for a few hours. Around 11:00 a.m., Alastair would go into the lobby area to watch an English Premier League soccer game while I curled up on the pool deck under the heavy shade of a tree to read. Pegasus was a sanctuary.

## Thanksgiving

Last weekend was the Canadian Thanksgiving. It was not a celebrated holiday in Guyana. There was no turkey, no cool fall air or brilliantly colored leaves, no hikes through a forest or fishing on a lake, no frolic and fun with loving family and friends. So, being cognizant that in Canada this was THE day to give thanks, we decided to make it a special day and to be especially thankful for everything and everyone in our life.

On Saturday we celebrated the fact we were alive and in good health, and as we sipped our morning coffee and munched on a fresh mango, we felt appreciative for having good shelter and food to eat.

Soon after 9:00 a.m. we headed out to the Farmer's Market. The variety of fruits and vegetables set out on vendors stalls seemed to be more abundant and fresher than ever. It was obviously much the same as last week, but with our heightened sense of gratitude and sharpened mind-set, the cornucopia of natural products looked more lavish and enticing. Our altered mind-set must have been apparent in our smiles for the vendors seemed to greet us with more warmth and bigger smiles than normal.

We then headed over to the Pegasus and lounged poolside. We luxuriated in the sun's heat while being silly practicing our Spanish – we were taking free Spanish classes at the Venezuelan Institute two nights a week, and getting a taxi to and from even though it was only four streets away from our apartment.

Later that night, to celebrate this special day in Canada, we went to an Indian restaurant for dinner with our Canadian volunteer friends, Ian and Marysia. Decorated with banners and walls adorned in dazzling gold and reds, with glittering baubles and bright lights, the restaurant was gaudier than the Indian restaurants we were familiar with in Canada. And to add authenticity, the waiters wore turbans and white coats, and the music was 100% "Bollywood."

As we tucked into our aromatic curry dishes, we shared favorite Thanksgiving stories from the past, and then changed the subject to International Volunteering. Ian and Marysia had spent time in Africa on a VSO assignment, and their humorous stories about life in little African villages elicited in Alastair and me a desire to go there one day. But, as we were now in our mid-sixties, we had to be realistic about life. Could we go that far away from family and for a prolonged period of time at this stage of our lives? We didn't think so. Too bad we weren't ten years younger. We all agreed it was regrettable we didn't discover the world of International Volunteering sooner; say in our forties, because it was such a meaningful and enjoyable way to live. We regaled our company with the Jamaican "bee" story and other humorous adventures from our volunteering history.

As the waiter cleared away the table, the conversation turned to the present day and the future. Though none of us had any firm plans as to what we'd do or where we'd go after Guyana, we were all of the same mind in that we'd like to continue to volunteer, somewhere in some fashion, if at all possible. We were thankful for the past events in our lives that had conspired to bring us together at that table, at that time. And with that deep philosophical thought, we parted company with full bellies, grateful hearts and energized minds.

Later that night, as we lay in bed reflecting and reminiscing about the people we'd met on our journeys, the places we'd seen, and the things we'd done, we felt great gratitude and joy in being so blessed with abundance and fortuity. It had been a special Thanksgiving Day we'd remember forever.

On Sunday, Carol, a British volunteer, had cable TV installed at her house and invited everyone over to watch a Premier League soccer game. Alastair headed over there (armed with his freshly baked French bread), while I caught up on emails and Skyped family members back home.

## On The Move Again

When Ann and Andy, fellow Cdn. Volunteers, were getting ready to leave the country, having completed their six month placement, we asked Cuso if we could move into their place. They agreed.

We moved in a month later and immediately enjoyed the comforts of a real home – a large living room with two cozy couches, a couple of soft chairs, coffee tables and lamps, an ample kitchen with restful chairs, and an open back porch with two hammocks. As Alastair put it ... It felt like a home not a motel! The location of this apartment was close to a large National Park where we could stroll through some green space on the weekends. It was idyllic. As we lounged in our hammocks and sipped fresh fruit smoothies, we thanked the Universe for listening to our plight and for sending us a home with good vibrations.

Though all aspects of the flat were wonderful, the best feature was the back porch. It was an open, grilled area overlooking lush trees and gardens, where parrots roosted for the night. They were noisy, but we didn't mind – we were at that point where any sound of nature was music to our ears. Actually, their racket made us laugh. It was as if they were bickering about where they would go for the day, before taking off every morning around 8:00 a.m., and we didn't see them again until dusk, which was around 6:00 p.m. Then it was as if they were excitedly talking about where they'd been for the day and what they'd done. This went on for about an hour, and then, just as suddenly as it began, it stopped and there was absolute quiet. Alastair made me smile because he'd call out from the porch, "The boys are back in town, the boys are back in town" when he saw them flying in. The back porch was also refreshing due to a constant cool sea breeze that wafted through the bars and across the hammocks slung across the porch. It was a little bit of heaven for us to start and end each day.

We shared this outdoor space and greenery with Lisa, a professional masseuse, who rented a room attached to the back porch. That was her office, and the porch was her waiting room for clients during the day while we were at work. How convenient was that! Lisa was well known, well reputed, and had a series of programs on women's health on the television. Of course, Alastair and I enjoyed many massages during our sojourn at this address.

Prior to meeting Lisa, I hadn't heard about breast massaging, but, after watching her television program, I decided to try it. I must admit I felt awkward at first, and though it was surprisingly painful, I do believe it is beneficial in that it releases toxins that gather in the mammary glands. I haven't heard of this massage therapy in Canada!

The second great aspect about our new home was our landlord, Norma. She was so welcoming and generous in ensuring we had all we needed. She lived on the main floor of the house: we lived on the second. She kept her home spotlessly clean; painted the concrete walls enclosing the property white; and lovingly tended to the flower pot gardens scattered throughout the enclosed court yard. Norma was a widower, having lost her husband to sickle cell anemia ten years earlier. They decided when they married not to have children because it was likely the disease would be inherited.

She loved her late husband so much that, though he'd passed over a decade ago, she continued to live her life in joy, basking in memories of their time together. Every Saturday morning, she played the songs she and her husband loved, and sang along. Many times we found her dancing by herself, but she obviously had her partner in mind, judging by the smile on her face and the position of her arms. We were mindful of her Saturday routine and respected her privacy. Sometimes, as we left the house to go grocery shopping, we'd glimpse her swishing and swaying on the patio, and it always softened my heart and brought a tear of joy to my eye.

Norma was also a fantastic cook and oftentimes surprised us with gifts of yummy dal, delectable curries, and heavenly roti. Although she said she didn't drink (it was against her religion), she said she sometimes would make an exception and have a beer. The exceptions became more of a habit on Friday nights. That's when she joined us and our volunteer friends in our apartment for an evening of social chit chat. Watching her have a beer —she only ever had just one – and sitting back, relaxing and laughing pleased Alastair and me greatly. We thought she particularly enjoyed the Friday nights because she sometimes asked mid-week if we were having friends over on Friday!

Norma did many things for us. She washed our clothes, cut Alastair's hair, sewed a button on his shirt, and was always most responsive when little things in the flat went wrong – so radically different from our previous landlord.

She was also religious. Around Christmas, after being invited many times to accompany her to the Jehovah Witness Church, we finally agreed to go. We were treated as honored guests. Everyone greeted us and introduced themselves. What was amazing was the confidence and poise the children and youth demonstrated. They approached us at different times over the evening, full of smiles and warm handshakes, ensuring we were comfortable and had everything we needed, which was of course a Bible. They seemed to sincerely want to know who we were and what we were doing in Georgetown. We thought the church's "teachings" must include the finer points of public discourse, as all congregation members made a special effort to greet us. We later learned their church did actively encouraged members to talk to strangers about what their church preached, and to spread the "Word." These young people impressed us greatly.

Since this apartment was quite far away from the NCD office, I took a mini bus to work. The buses were independently owned and were all competing, quite dangerously, for each passenger standing at the bus stop. Many times, the buses would come, one right after another, and would race and jockey down the road against each other to get to the bus stop first. Remember, there were few traffic lights, so honking represented the right of way at intersections.

I began to walk home from work in the afternoons when Alastair and I didn't go to the Pegasus, more for exercise, than to look at the scenery on route. Georgetown was unsafe for pedestrians, not just because of the chance of being robbed, but because of the lack of infrastructure. There were few traffic lights and no sidewalks. One had to be mindful at all times that cars, buses and trucks didn't care about the safety of pedestrians. I never carried any valuables – only a few dollars tucked inside my bra in case I wanted to pick up something at the market.

Because my laptop was my lifeline to family and friends back home, I refused to carry it back and forth to work for fear of losing it to a thief. This created a bit of a problem because my work place didn't have a computer for me. I had to use the one in the library, which was somewhat disruptive to my work flow, but I managed.

## Preparing for Trouble

Guyana held an election in November. There were two main political parties, each representing a different ethnic group: the Afro-Guyanese and the Indo-Guyanese. In the past, the vote count from these two groups had been close, and so the challenge for each party was to woo the Amerindian vote and thereby win by a majority. There was a history of things getting pretty heated up around elections, but it was hoped those violent times were in the past. Just to be on the safe side, Cuso advised us to stock up on extra supplies in case we had to stay at home for a prolonged period. They also had an evacuation plan in case the situation got totally out of hand. We weren't too concerned though as there hadn't yet been any reports of pressure or tension in Georgetown. We went about our business in a normal, quiet and unobtrusive manner, using common sense and purposely avoided engaging in any political discussions with work colleagues.

On voting day, all volunteers were put on alert. We had to stay at home, or as close to home as possible, and to have our cell phones available to receive a call from Cuso. They designated our house a "safe house," and stocked it with extra bottles of water, food rations, and bedding. If the post-election climate moved from green to yellow, four volunteers would take refuge with us. And if a state of emergency was to be declared, we'd all be flown from Georgetown to Letham, and if needed, on to Brazil. That didn't happen, but Cuso had taken all precautions to keep us safe and out of harm's way.

## The Work Continues In spite of Obstacles

International Development is all about building the sustainable capacity of people and organizations. We know when we arrive at our placements that more often than not our work plan will change. However, I was surprised when my Guyanese co-worker left two weeks after my arrival, on sick leave, which was then followed by an education leave of four weeks. This meant I wasn't supporting a co-worker's capacity to develop the NVDP, but was actually doing the job myself! The pressure to develop this program without the in-country knowledge of my co-worker was vexing. I was mired down trying to piece together statistics and outcomes from the past to apply for funding for the future. I was also responsible for completing the first funding report for matters that took place prior to my arrival. As frustrating as it was, I'm pleased to say I completed and submitted the report in a timely fashion.

What I enjoyed most about my placement was the time I spent with the three volunteers who were the program facilitators. They were amazingly resilient individuals who'd dealt with and had overcome unbelievable obstacles in their lives.

Alastair's placement was continuing to be challenging because of the YLA members were not usually available to meet with him. The Strategic plan was done and he was now working on fund-raising. He met with his Cuso program director who suggested he continue working with the youth when he could, and act on their behalf in meetings with local leaders and persons of interest in the community. He continued to send out letters and funding applications.

## Kaiteur Falls

The Kaiteur National Park, established in 1929, encompasses 627 square kilometers of pristine jungle, creeks and rivers teeming with biodiversity. Kaiteur Falls is the world's third longest single drop waterfall – four times higher than Niagara Falls and twice as high as Victoria Falls in Africa. It is also one of the most powerful waterfalls in the world with an average flow rate of 136,200 liters of tannin-stained water per second.

Legend has it "Kaie," a great Patamona chief, sacrificed himself by paddling his canoe over the falls to appease Makonaima, the Great Spirit, to bring peace between his people and the aggressive Caribs. "Teur" translates as "Falls," hence the name, Kaieteur.

We hiked it, but had Fiona, a twenty nine year old Irish volunteer, not convinced us – she said it was an easy hike – we wouldn't have done it. Her parents were visiting from Ireland and she wanted us to join them on a three day trek up to the top. It was grueling and exhausting and took every gram of energy we had, but we persevered and made it, and were ever so pleased we did. There was no way to be airlifted out on route, no matter what happened, because it was all undeveloped jungles, rivers and mountains. What we saw and experienced was wonderful on so many levels.

There were seven of us: Halina and Steve from the UK – she was a VSO volunteer and Steve was her accompanying partner; Philippa and Peter from Ireland (Fiona's parents); Monica, another UK VSO volunteer; and Alastair and I. As we were all in our 50's and 60's, it was a well matched group. Fiona had done the hike earlier and wanted her parents to do it, but she didn't have enough vacation time left to accompany them. There needed to be exactly seven people for the trek to be a go, and she needed five more. Was that why she said it was an easy hike? Anyway, the upshot was we, without exception, were glad Fiona had conned us into it.

One back-pack containing a change of clothes and enough snacks and water for a full two days plus a hammock was all we were allowed to take.

Two Amerindian guides, Tony and Alwin, were to escort us on our trek: one to lead and one to be last to ensure no one was inadvertently left behind. That was reassuring.

We all met up at the Stabroek Market in downtown Georgetown at 6 a.m. one Friday morning, except for Alwin who would join us later, to board a mini-bus to Mahdia. Along with the passengers, there was also commercial cargo: boxes, bags, a television and even a car transmission, which were stored wherever there was room. Some items were strapped on top, some underneath, and some crammed into the back part of the vehicle.

Except for a few towns, such as Linden and Mahdia, the interior of Guyana is sparsely populated and the terrain is basically undeveloped rain forest. All roads out of the coastal strip into the interior dramatically change their composition when they get out of town: from paved highway to loose gravel and then to a wide, pink, dusty dirt track. When it rains, the dirt morphs into thick mud, and then, if the rain persists, which it often does because Guyana is in the tropics, the road floods and rivulets with fast flowing, waters appear. Many vehicles running those roads have watertight undercarriages and vertical mufflers that look like periscopes. We didn't have that luxury.

To combat illegal immigration in the country, especially from Brazil, the police set up check points at police stations along the way. We had to go into the station at Linden and Mahdia to prove our identity by showing our passports before the police would allow us to continue our journey. In Mahdia, as we waited in line, I saw a prisoner peeking through an iron grill on the upper part of a steel door, located directly behind the reception counter. He looked at me with hungry eyes, and I smiled, because the sign on the wall next to the grill read: Feeding times: 9:00 a.m. and 5:00 p.m. I questioned the meaning. Did it mean visitors were expected to feed prisoners, but only twice a day and at certain times? No wonder the man looked hungry!

It was a long, bumpy ride to Mahdia, but we didn't mind the van's discomfort, nor were we overly concerned when it swayed from side to side through muddy sections. Every sight and sound, as we drove through small settlements and countryside, was novel and thrilling.

We arrived in Mahdia, the gold mining center for the region, about ten o'clock. The place had the feel of a Wild West frontier town: what Dawson City must have looked and felt like in the 1890 gold rush era. It was full of prospectors looking to buy supplies or to sell their gold, and was well equipped with bars and the other establishments that take root in such a town. We had to laugh when we saw our pick-up driver because he had a mouthful of gold capped teeth. He'd panned the gold and kept his wealth in his mouth for safe keeping! We saw many people around town with gold teeth. They purposely smiled a lot – to show off their good fortune no doubt.

From Mahdia, we ferried up the Essequibo River to Mabura, and then piled into the back of a pick-up truck with our entire luggage for the rough ride down to the Potaro River. We crammed into a small wooden boat with a 25 h.p. motor to go to Amatuk, an island in the river, where Alwin would join us. We were now plying the waters of the Amazonian rain forest basin.

The Essequibo and Potaro rivers were stunningly tranquil and pristine. The water was brown due to silt, a trademark of rivers in South America, and the surface was as smooth as glass. But underneath, Tony said, there was a strong under-current making swimming dangerous. What a pity! It looked so inviting. Here and there, we saw gold dredgers, and on the shore, small mining camps, where locals panned for gold. At $1,800 an ounce, it was an exceedingly profitable endeavor.

We made one quick stop on the river bank to drop off a miner's wife and her provisions. Many miners independently panned for gold on the river's tributaries, and spent months alone doing it.

It was dusk when we arrived at Amatuk and met Alwin. Tony suggested we refresh ourselves in the river, but not to pee in it, because there was a bug that would follow the pee stream up into your body and cause you great discomfort. But, he kindly reassured us, there were no piranhas in this part of the river. That was particularly good news because a couple of months earlier, a piranha chewed the toe off a volunteer in Georgetown when he swam in black water. He returned to the UK for treatment, but I'm pleased to say, he was back in Guyana continuing his work, and showing off his "war wound" to the curious.

I did dip into the water before breakfast the next morning, but stayed close to the edge. I couldn't come this far in the wild, untouched rain forest of Guyana without christening the water with my presence. It made being there that much more real for me.

We climbed up from a small beach to a flat pad of ground with a binap (hut with no walls), and hung our hammocks for the night. What a laugh! As it was a new experience for all of us, we didn't know how high to hang the hammocks, nor how long to stretch them. As it was, we hung them too close to each other, but then again, we only had so much room to work with. When one of us turned over in the night, we sent abutting hammocks swaying and they in turn swayed the hammocks on either side. And of course, being an elderly group, some of us seniors had to get up in the night for a pee.

Alwin reminded us, as we prepared to retire for the night, that we were in a jungle and should expect to see anything: jaguar, large spiders, snakes or huge insects. He said not to venture far from the binap and to check our boots before we put them on because scorpions and other nasty insects with stingers liked dark, smelly places! It was fortunate there were a full moon and a clear sky because it was light enough to see without a flashlight (which of course we didn't have). All things considered, most of us slept okay, but there were reports at breakfast about snorting, snoring and other nightly sounds that the insomniacs joked about.

The following morning, we hiked down to the river's edge to board a different boat. Because this boat was smaller, Alwin had to make two trips down the Potaro River. Alastair and I were on the first one along with Monica and everyone's luggage. It was a long, memorable ride – about an hour – before we reached the rapids. I sat up front on a bench with Monica, and Alastair sat in the back with Alwin and the motor. I've never had such a thrilling experience and in such an exotic setting. It was wondrous. Alwin dropped us off at a portage area and returned to get the next party.

We hiked inland a short distance to the Ranger Monitoring Station and met Orel, the ranger, who just happened to be Alwin's brother. He made us coffee and answered our questions about living in such an isolated location. It was enlightening.

Although we (Alastair, Monica and I) had bright sunshiny weather for our trip down the river, the next group wasn't so lucky – they arrived drenched from head to toe. The heavens had opened and released the cats and dogs. They wrung out their clothes as best they could, but there was no time to waste, we had to keep moving.

We hiked around a bend to circumnavigate the rapids and to our next boat. Alastair's response to what we found was, "You've gotta be kidding!" The vessel that was to carry all nine of us and our bags was submerged and bits of the boat were letting go. Discretely, I asked the others how they felt about traveling in this tub. They too were concerned, but what were we to do? We had no choice, but to trust the judgment of our guides. Surely they wouldn't put us in harm's way?

Tony spent the next fifteen minutes bailing out the boat. We then loaded it up with our gear and piled in. Steve and Peter sat at the back and bailed for the entire journey, as water continued to pour in from a hole at the front. Now and again, the motor sputtered and quit, but Alwin always managed to get it going. We were relieved and grateful for his skill for there would be no help available on such an isolated river.

Puttering slowly down the Potaro was enchanting and a memorable experience. The continuous solid green of the jungle on the river banks, mile after mile, was mesmerizing; the still, black water with its reflected trees, sky and clouds was stunning; and the cool water that occasionally splashed up and cooled us was delightfully refreshing.

An hour later - it was about noon by then - we rounded a river bend and there, framed between two mountains was Kaiteur Falls. What a sight! All this greenery split in two by a sheet of white from blue sky above to brown river below. It was an awesome vista. And to think we'd come to actually hike up there seemed absolutely preposterous – an impossible feat for us oldies, surely!

The ride down the river had been so exciting and remarkable that it alone was worth the effort we'd invested up until then. And had we gone home at that point, we'd have been most satisfied. But that of course was just the start of the adventure. As we neared the base of the falls, we saw and felt the force of the water plummeting and roaring down into the river, creating a massive cascading spray.

Once on the river bank, we rested for a few minutes, but couldn't tarry long because we only had so many hours to reach the summit. Never in my wildest dreams did I ever think that one day I would have this kind of adventure.

It was about noon when we set off and about five when we reached the top – so much for it being an easy two to three hour hike. It physically demanded every ounce of energy and every gram of muscle we had, and we all agreed it was the most physically demanding feat we'd ever done in our lives. For Alastair, he said it was even more demanding than climbing up the two kilometer stretch from the valley in Cherrapunjee, India after visiting the living-root bridges, and he thought at the time there could never be anything remotely more strenuous than that. Just goes to show we should never say "never."

For five hours, we climbed up, always up, and over stones, rocks, streams, roots and fallen trees. After a few hours that seemed like days, the thought of when we set off seemed weeks ago. Tony led the group and Alwin positioned himself last to ensure none of us got lost.

We chose to be last because we wanted to see and savor everything on the way and not feel rushed. But there wasn't much to see. It was just trees, vines and rocks. The most exciting life form we came across was howler monkeys. They looked down inquisitively at us as if to say, "What are you? Why are you here? Are you edible?" Once in a while, the bright flash of a blue morpho butterfly darted through the woodland. That was magical.

Being in the dense jungle, soaking wet from perspiration brought on from extreme exertion and high humidity, was exhilarating. With every step we were making progress up the mountain.

The others charged ahead and we sensed they were a little disappointed when Tony stopped them periodically to allow us to catch up. Perhaps they'd set a personal goal of how quickly they could make it to the top, or maybe they were just in better shape than us. Regardless, we were determined to stay present in the moment and enjoy the reality of where we were and what we were doing. When we weren't puffing and clambering, we spoke with Alwin and learned much about him, his family and life in that part of the world.

After two and half hours, we stopped and gathered at a clearing for a rest and refreshment and to chat about the experience. We were exhausted and laboring, but we all felt we must be getting close to the top. We just needed a break to fuel up for the final push. Then Tony shocked us with the dire news that we'd reached the half-way point!

Crazy thoughts ran through our heads – we had hammocks in our back-packs – maybe we could call it a day, string up a hammock for the night and climb the rest in the morning. We'd be okay because our plane wasn't expected until noon. We knew we couldn't go back because there was no reverse transportation available.

"No, you have to finish the climb today," Tony said with a hearty laugh. And once he saw that we'd accepted this reality, he stunned us with a further piece of news: "You haven't climbed the three OMGs yet!"

We looked alarmingly at one another.

"The OMGS? What on earth are they?" we all chimed in.

"They're the "OH MY GOD" sections!" Tony replied, and roared with laughter. You could tell he enjoyed making this statement and probably had been looking forward to it for the last hour.

"Don't worry. I'll be right behind to push you up if you can't make it," said Alwin, and both he and Tony slapped hands and guffawed.

It was about this time I changed my mind about getting wet. Up until then, I'd been most careful to step on stones and avoid getting my running shoes wet, but now I didn't care – it was more trouble than it was worth. In fact, the opportunity to slosh through streams to refresh myself was most welcome.

What apt descriptions "OMG" was. This was where the trail became more vertical and we had to literally clamber up over stony, wet rubble with our hands and feet in tandem. We pushed and pulled one another and sometimes had to claw our way forward. Our calf muscles, knees, thighs and biceps strained with the effort, but in a strange way, we all agreed it felt good – how weird was that? Maybe it was the sense of achievement we got from doing it, or maybe from relief in the knowledge it was now behind us. And don't forget, we were each carrying a heavy back-pack with clothes, shoes, hammock, bug net, snacks and water.

At one spot, spring water gushed out of a ledge in the rock. We didn't undress, but we bathed in it as totally as we could. It was wonderful to be cool all over, even if the feeling only lasted a few minutes, before body heat warmed it up and changed it into perspiration and then to steam.

After an hour and twenty minutes we reached the studio, as Tony called it: a flat area with a bench made out of a log. We'd reached the top of the mountain. Hallelujah. Tony chopped up a pineapple to celebrate the hard won milestone. There was still forty minutes of tough slogging to go though, before we reached the Falls. The terrain from this point on was flatter, but when you've expended every ounce of energy you have, every footstep is hard work. Eventually, we walked out of the trees and there, off in the distance to our right, was Kaiteur Falls. They were magnificent. And what I found most surprising was they were quieter than I had expected: so unlike the roar at the top of Niagara Falls. The reason for this was because the water went over the top silently, and only thundered when it roared down and smashed into the river nine hundred feet below.

We proudly took photos of one another with the Falls in the background, and continued on to other viewing spots as we got closer and closer. We already knew about the giant bromeliads, some up to ten feet high that grew profusely in that area, and so weren't surprised to come across them. But what was absolutely astonishing was when Steve looked into one and spotted a tiny – no more than a half inch long – endemic Kaiteur golden frog, also known as the "poison dart frog." It was just sitting there waiting to have its photo taken, in all its brilliant yellowy orange finery, set against a backdrop of a thick, deep green leaf. It spends its entire life in the bromeliad. How lucky was Steve? Many people search fruitlessly for days and even weeks to find one and some never do. The frog benefits from the good hiding places among the leaves and from pools of water collecting at the base. The poison comes from eating insects, which have in turn, eaten certain plants containing toxins. Indigenous people collect the poison from the slime that oozes from the frog's skin to use on blowgun darts and arrows for hunting. It sure was our lucky day and what a gift for our troubles.

The Kaiteur Falls National Park Lodge was quite a humble affair – a simple wooden building with just two bedrooms, two washrooms, a kitchen, and a covered area to hang hammocks. Only one bedroom was available – two guys from Holland were in the other one. I drew straws with Halina and won, so Alastair and I got to sleep in a real bed that night.

Tony went off to buy two bottles of rum, some coca cola and juice from a small village somewhere in the rain Forest, while Alwin cooked a delicious vegetarian supper. We were all astonished that somewhere in the middle of the Amazon Jungle was a shop selling Coca-Cola and rum!!

Tony returned two hours later and we celebrated in true Guyanan style, with five year old Demerara rum and coke. Everyone hit the sack shortly after, except for Tony, Alwin and Alastair. Though he was "knackered," Alastair somehow managed to get his second wind – the rum and coke helped I'm sure – and they entered into a rich, lengthy discussion about aboriginal rights and governments. He thought he was having a private discussion, but the next morning all told they had heard every word because the men were talking so loud. No doubt the contents of the glass had something to do with it, but knowing Alastair, it would have been contents of the discussion – the opinions and views of the three guys – that would have produced the hearty dialogue.

Everyone got up at 6:00 a.m. and drank instant black coffees. Alwin cooked a tasty dish of stewed vegetables and fried a confection he called "floats" – somewhat like a roti or a poori. You open it up and spoon in the hot vegetable mix. It was substantial and tasty and just what we needed to satisfy our ravishing appetites.

Once we packed up and were ready for the plane, we headed out to further explore the area. We found a path, close to the Lodge, that led right to the Falls – one could step into the water and dive off the top if one wished! We did step out, but only to peer straight down over the edge and marvel at Potaro's power, thundering free fall down to the misty turbulence at the base. While gazing at the pristine rain forest valley in the distance, I experienced a sense of vertigo mixed with awe.

About twenty feet from the edge of the Falls was a little eddy with swirling waters, and a local woman was there bathing her child in it. When she began swirling him around to get off the soap suds, our hearts jumped into our mouths, and our brains cried out, "Oh My God. Hang on to the little guy." We watched frantically as she went about her business, and then realized she'd probably done this with all of her children every day of their lives.

It was a different eco climate up there on the mountain top from the jungle of the day before in that it was densely covered in mist from the Falls, and was cool and refreshing. At times, visibility was no more than ten feet. We set off in search of the Guyanan Cock–of–the-rock bird: a beautiful rare orange species with a large crest on its head. It's about twelve inches long, weighs about half a pound, and eats mostly fruit, but sometimes, if it's starving, it will eat small snakes and lizards. We were close to where they nested, but unfortunately we weren't as lucky as the day before because Tony and Alwin came to get us. The eight-seater plane had arrived to take us back to Georgetown.

The flight was about an hour and a half – ninety minutes of peering down on pristine jungle and rain forest. For eighty minutes, there was no visible break in the scenery except for when the Essequibo and Demerara Rivers came into view. It was pure greenery all the way – no sign of any settlements or cuts to indicate roads, other than the one we traveled on to Mahdia. What an amazing country! I could now see why it was a carbon sink for the world and why it should be preserved forever as it is today. Guyana is a unique, remarkable and undeveloped country I hope never changes.

*Kaiteur Falls*

*Al and Candas on Kaiteur Falls*

*Candas and bromeliads*

## Christmas Holidays

Although we missed our family terribly and were eager to see them again, the thought of heading back into the dead of winter to fly to Toronto, and then embark on a two hour car ride to London, Ontario, possibly in a snow storm, wasn't something we wanted to do. Clothing was another consideration because we left Canada in August and didn't bring any winter clothes with us.

One thing for sure though was we both felt we needed to get out of Georgetown for a change of scenery. What I wanted was time to be still; to listen to the beat of my heart; and to center myself in just being. I wanted to recharge my batteries and return to Guyana with renewed energy to serve the community I was working in. This was important to me.

## Tobago
### (by Alastair)

Volunteering is work when all is said and done, and can be quite stressful too, because one is working in a different culture and in a different work environment – usually with fewer resources than accustomed to having back home. Also, one bears the responsibility for achieving stated results. It's most important therefore for volunteers to periodically take time off and have a change of scenery to recharge their batteries. The upside to this is the volunteer is usually in a location offering numerous possibilities for holidays in areas they'd normally never get to in mainstream life (unless they were millionaires of course).

When I was in Bangladesh I went to India three times on vacation, and now we were living on the Caribbean coast with dozens of exotic islands to choose from. How sweet was that! The full reality of how fortunate we were truly hit home when we researched which island to go to. There were so many to choose from and all were doable in terms of cost. It was pleasurably overwhelming.

After much research, we decided to go to Tobago for ten days over Christmas and New Year. It was the closest island to Georgetown – less than an hour flight away – and seemed to offer the beauty and tranquility we were looking for. Imagine our delight when we found a last minute special on airfare of $45 round trip per person. And if that wasn't fortunate enough, we then found an amazing yet economical place to stay. Millers Guest House was a relatively small complex of cabins, dormitories and guest houses located right on the shoreline at Buccoo Bay, a quiet and relatively remote part of the island with few visitors.

Our little flat had a bedroom, living room, washroom, kitchen and a small balcony that overlooked the bay. To our right was a small pier where local fisherman docked every day to unload their catch. Twice we bought fish there and cooked it in our apartment. One fish that was memorably delicious was bonito, a type of dark tuna fish. The kitchen was well equipped with everything needed to make life comfortable – oven, microwave, blender, toaster, coffee percolator and a full set of knives for filleting fish etc. And we even had hot water – what a delight to have hot showers again.

Ninety percent of the time we were the only ones on the beach and that surprised us because the sand was soft and clean, the water gentle and refreshing, and the ambience idyllic – perfect for swimming or just bobbing in the gently lapping waves, whilst savoring the beauty of the sky and surroundings. After breakfast, we sometimes walked along the cliff path and were awed by the scenic tranquility and beauty as we gazed over the cliffs and out to sea.

Next to our accommodations was the El Pescador, a great restaurant run by Leonardo, a Columbian. He put on dinners at Christmas and New Year for the eight guests of Millers Guest House – turkey, ham, roast beef and chicken with all of the usual festive fixings.

Not much happened in Buccoo Bay on a daily basis and that was the reason why tourists didn't flock there mid-week – no commissions to be made by tourist agents in Scarboro, the island capital. Sunday evenings were the exception. Day-trippers from various resorts on the island bussed in to Buccoo around supper time for an event called "Sunday School" – a fifteen piece steel pan orchestra played all genres of music, from symphonic pieces to Broadway tunes to island calypsos. It was noisy of course with the bands and the crowds, and the smells from meat cooking over open fires ignited our appetites. What a hoot it was. Everyone was in a festive mood and danced and smiled. The music compelled you to move your feet, sway your hips and laugh. And did we ever laugh. It was wonderful to be in love and to be in Tobago.

They call Tobago "Paradise Island" and I can vouch for that. It truly was a fabulous place: picturesque bays, warm seas with great snorkeling coral, and an easy going people who liked to talk and help. In our travels around the island, and we traveled extensively to all parts, we were surprised at how few poor areas we encountered, and how little industrial activity we saw. Tobago was so unlike its neighbor, Trinidad, which was bigger and more diversified. Everywhere we visited was well kept, litter free and freshly painted. The unemployed went to Trinidad for work resulting in a population base in Tobago being comprised mainly of older family members living traditional lifestyles in little villages, local fishermen, boat builders, as well as people in the tourist sector – shopkeepers, restaurateurs, resort staff etc. Never have I felt so safe and relaxed, and that's quite a statement because I thought Costa Rica was tops in that category.

We made most meals at home, as was our custom and preference. But we did eat out twice at a little Italian Restaurant in Buccoo, because it had such an inviting ambiance from the outside. We just had to go in and try it out, and we're so pleased we did. It was romantic, being lit only by candlelight and the food was to die for –the owner and chef was Italian and it was obvious he knew what he was doing in the kitchen. His linguine and cannelloni were sensational. He proudly informed us that the reason his food was so authentic and tasty was because he imported many ingredients – olives, cheese etc. directly from family members in Italy.

Most days began on the beach with a dip and a sun-bathe. Around eleven, we made tracks and went off somewhere for the day. We spoke to many people: boat builders, taxi-cab drivers, establishment proprietors, as well as tourists from all over the world. We met many interesting characters, young and old, and many people who inspired us to be different by their non-mainstream lifestyles and attitudes. One such individual was a fellow Canadian, only about fifty years old I'd say, who was living permanently at Millers Guest House. He had a shipping container full of boat parts that he intended to unpack – one day. He didn't seem to be in any hurry, but there was a hold-up on the paperwork he said! Another full time resident at Millers was a Russian who had a passion for photography.

Buccoo has a goat-race track. That was odd, because it was only used twice a year, yet it was a substantial real estate holding and well maintained. January 2, a Sunday, was Family Day in Tobago, and one of two days in the year when races were held – again, how lucky for us! Steve and Halina, two of our British volunteer friends in Georgetown, were also on the island and came to visit us as a special treat because it was Steve's birthday and they wanted to do something different.

Crowds began arriving at the track around eleven, and by noon, when the first race started, there must have been at least two hundred people in attendance – not many in Canadian terms for a special event, but a lot for little Buccoo Bay. It was party time. We picked out our goats after watching them parade around the paddock area and placed our bets. None of us won – we'd somehow all managed to pick the losers: the last ones to come across the finish line, and even some who didn't make it that far! By the third race, we realized what we were doing wrong. We were appraising the goats ability whereas we should have been looking at the goat handlers, for they invariably ran faster than the goats. We changed strategies and had better results, but not enough winnings to be able to retire in Tobago.

Although I'd vacationed previously in the Caribbean, it had been at all-inclusive resorts with little opportunity to truly savor what the islands were all about. Being free to come and go and explore whatever took our fancy was a new experience and one that suited us perfectly.

Buccoo Bay was a thirty minute taxi ride into Scarboro, the island capital, where the bus station for buses to all parts of the island was located. Getting a taxi was cheap (about $2 Cdn) and also easy because the taxis on the Scarboro to Buccoo route waited on the main road in Buccoo, about two minutes from Millers, and at a specific location in Scarboro. Every other day, around mid-morning, we took a taxi into town and connected with an island bus to take us to a different beach. After a few hours of sun and sand, we left the beach, walked up the road and over a hill to the next bay, and spent another few hours lazing, snorkeling or sleeping. Sometimes there was a bus stop near the beach, but not always. In those cases, we simply stuck out our thumbs and hitched a ride to the next bus stop or into town if the driver was going that way.

One particularly memorable ride was with a couple in their late teens, I would guess, with an infant. Candas got into the back seat where a young mother was nursing her baby and I sat in the front. With reggae music blaring away from oversized speakers, the driver toked on gangi and sang along, whilst the smoke blew into the back seat. And then the driver offered the cheroot to me and Candas. We declined. He smiled and handed the "smoke" to the young lady who was nursing. She was one mighty happy mother I can tell you.

By the end of our stay, we'd visited and savored all coasts and covered most of the island's interior. Our particular favorite beaches were Charlotteville and Pirates Bay up in the north west of the island and Castares on the north shore. The buses were comfortable and every journey was a delightful experience in itself.

Tobago was everything we dreamed it to be, and we highly recommend it to those tourists who enjoy moving around and exploring roads less travelled.

We returned to work refreshed, happy and ready to go at it until our next vacation.

# Guyana
### continued (by Candas)

## A New Year

After a bit of a rough start, we learned to live and function quite comfortably in Guyana. Sometimes accepting "what is" is difficult, and one has to let go of expectations, and focus instead on doing your best with what you have. After the holidays, our social life became unusually busy. There were parties most weekends for departing volunteers and celebrations to welcome new arrivals. The parties were usually held at a volunteer's house with everyone bringing their own food and drink.

The Filipino volunteers were a light hearted group, who loved music, particularly Karaoke. They had a software program, speakers and a microphone, so whenever we all got together, you could be sure we'd be singing the hits of the sixties and seventies. For some strange reason, everyone, no matter what country they were from, seemed to always want to sing the Eagle's "Hotel California" and all of ABBA's greatest hits. Filipinos also liked to cook and so there was no shortage of rice and fish dishes to sample.

## Volunteer Workshop

In February, Cuso facilitated a three day Volunteer Workshop at a local hotel. All forty volunteers attended and it was pleasing to finally meet those who worked in remote villages. It was a participatory, fun and interactive event. Volunteers spoke about their projects, what they'd learned, and shared tips, ideas and suggestions. Cuso outlined their development plans for continued work in Guyana.

## The Light at the End of the Tunnel

My co-worker returned after her eight week leave and with the three PWD volunteers, we formed a team to develop a National Volunteer Disability Program (NVDP). Things really geared up in January and February. We facilitated fourteen Sensitization Workshops to make people aware of the Act, including teachers, nurses, police officers, parents, other PWDs and social service organization workers. The workshops focused on engaging individuals in group activities, and in role playing to make them aware of PWD rights, as well as the NCD's role. Our work focused mainly in Georgetown where we sensitized about six hundred individuals,

45% of whom said they were unaware of the Act or the rights of PWDs prior to the Sensitization Session.

In between providing Sensitization Sessions, the three volunteers worked on other specific tasks. Simone's topic was accessibility issues. One of the highlights was the installation of the first "Accessibility Parking" sign in the whole of Guyana, in front of the main Public Post Office. This took six months to happen, but in the end we were gratified to have an official ribbon cutting event, officiated by a Government Official, and to get good coverage in the local newspaper. You have no idea how many government departments had to approve this sign installation, and crazy as it may seem, the only way it was ever going to happen was because NCD paid all the costs. Going forward, the volunteers will continue to work with the Ministry of Housing and Labor to implement Accessibility Parking signs at all post offices. It was a small, but significant step in making the government accountable for implementing its own laws.

Another key element of the program was an outreach activity facilitated at the public hospitals. We usually spent a couple of hours once a week talking to the people attending clinics. The hospital scene was unlike anything I'd ever experienced. The buildings were old and decrepit, and the process of attending to the sick was extremely slow and cumbersome. Simone and I arrived about 7:30 in the morning. She made an announcement to the crowd, which numbered about seventy souls waiting to register at the Clinic Attending Office, and then presented an overview of the NCD. She explained how the Commission could assist PWDs to access Disability Financial Public Assistance, Rehabilitation Services, accessible transportation to the clinics and access to aids, such as wheel chairs, white canes etc. During our hospital clinic visits, we helped many people complete the National Registry form, as many could not read or write or were blind and required someone to fill in the form for them.

I was amazed by how many people had permanent disabilities in this country: a large number were blind due to a lack of treatment for Glaucoma. The last census taking in 2001 listed 50,000 as the number of people with a permanent disability, however, in discussions with medical professionals and police officers, they estimated the real number to be much, much higher. More recently, many permanent injuries were being caused by industrial and traffic accidents.

Ganesh worked on Communication. He engaged government ministers to arrange workshops for civil servants. He also worked on the development of television and radio broadcasts and interviews and even hosted a weekly talk show.

Renatta focused her work on the National Registry and responding to the needs of PWDs for Rehabilitative Services. It was a major task to manage the distribution of registration forms to all parishes across Guyana for the registration of PWDs. The completed returned forms had to be entered into a database, which eventually required the assistance of Dennis, another volunteer.

## Homemade Bread and Other Yummy Food

Alastair completed his work plan with YLA and met with the Program Manager to discuss the status of his placement. They suggested he continue his endeavors to engage the youth and the Agricola community, in whatever capacity he could, and this he did, but he had a lot of free time. What to do? Hmmm.

He put his energy and passion into cooking wonderful meals, such as French bread, Indian curries, samosas, pakoras and making wonderful combinations of fruit juices. This was fine with me, and I had the weight gain to prove how much I appreciated his culinary talents!

He usually walked to the Cuso office when he had business there, both for the exercise and for something to do - he had a lot of time on his hands. Onetime, after it had rained heavily and the city had flooded, he had an interesting escapade – he fell into a hole, filled with stagnant water, up to his waist.

As he tells it, he thought he was walking in a grassy area of the street, when all of a sudden; he disappeared from the street and found himself in a four foot deep hole. He was pleased his hands had reached out and grabbed the sides preventing him from being totally immersed in the sludge, but even so, he was sodden from the waist down. He climbed out, and trying to come to grips with what had just happened, took off his wet socks and shoes, and wondered what he was going to do. No cabbie would want him in their vehicle, and would he be okay to walk home barefoot? He looked around and saw a home owner laughing and motioning for him to come over. The compassionate man took Alastair into his back yard and hosed him down with the garden hose. This, he said, made him feel better – knowing there was less slime on his body – but he was no less wet. He flushed his shoes, socks and pants legs, and hurried home as fast as his sloshy sandals would allow because he couldn't wait to get into the shower, get dry and put on fresh clothes.

He was fortunate he wasn't hurt because other volunteers, who had similar experiences of falling into holes in the wet season, had suffered injuries, such as broken arms and gashed foreheads.

One volunteer laughingly said she once drove her bicycle into a hole too – just the once mind you, because she now refused to cycle in the wet season.

## Mashramani

We experienced our first Mashramani Festival (Mash) celebrating National Republic Day in Guyana in the first week of February. The week-long celebration included dances, parties, the making of decorative costumes and floats to partake in a huge parade much like Carnival or Carabana. The parade lasted about six hours, winding up in the National Park with the float judging and lots of music and dancing. For the Guyanese, it was the biggest celebration of the year and after the parade, street parties continued way into the wee hours. Many people returned home from overseas specifically to celebrate Mashramani.

We went to the parade, watched for about an hour and a half, and then returned home to our sanctuary, away from the sun, the noise and the pollution. You wouldn't believe the size of some of the speakers on street corners and on floats. Some must have been at least twelve feet high and as wide, and the decibel level emanating from them was deafening.

The parade route was littered with an unbelievable amount of garbage. Had no one looked for a trash bin? Perhaps there were none?

Cleaners began to clean up as soon as the parade finished and even though they worked throughout the night, there was so much of it that it took another two days to get the job done.

The trash in the canals, trenches and gutters tripled during the days before and after Mash, and even though they tried to get it cleaned up as fast as they could, the rains came before they could finish. Georgetown and hundreds of homes flooded. I wore Wellingtons (rubber boots) to work as the water on the streets was about three inches from the top of my boot. The water was also polluted. Continuous health warnings on TV and radio, broadcast every fifteen minutes or so, reminded people to soak their feet in water with bleach to cleanse them, if they had walked through the street water.

It was cooler now in the evenings and early mornings, which was refreshing, but it heated up by mid-day to quite high temperatures. It was the so called rainy season and by God did it rain. I've never experienced rain coming down with such volume and velocity – you'd swear someone up there was dumping bath tubs of water down on you, and it was incessant. The streets flooded and made a mess of your shoes if you didn't have boots. It was dry when I left for work one morning and so I wore my sandals. It rained all afternoon, and by five, when I set off for home, the streets were flooded. I took off my sandals, rolled up my pant bottoms, and waded home barefoot. I washed my feet well when I got home, but not with bleach I might add. This happened on a number of occasions. I really needed two pairs of rubbers: one to be left at home and one to be left at work.

The mosquitoes were annoying and ever present, so one had to always have a layer of repellent spray on all exposed flesh or else you'd get bitten alive. We got an electric tennis racket, courtesy of a returning volunteer, to smack them with. Funny how past skills can be used in innovative ways when needed. Alastair used his tennis strokes: serves, volleys, forearm smashes and lobs, to effectively dispose of the little biters.

## GO FOR IT

### Travel Is Difficult in Guyana

We didn't travel much outside of Georgetown because it was such a chore and most unpleasant. One had to travel for long hours, in an overcrowded mini bus, with a very scary driver on badly washed out rutted roads to get anywhere. The only other way to travel was by plane, but that was too expensive for us frugal pensioner volunteers on a budget. There were so few guest houses or hotel rooms in the villages and small towns, and they were surprisingly expensive. Guyanans stayed with family or friends of friends when they went out of town, and so the guest houses were mostly used by business people on expense accounts.

We'd met the rural volunteers at the Cuso workshop, and so had a connection with some people who we might be able to stay with if we traveled out of the city. We considered going on holiday to Lethem, on the border with Brazil, before we left Guyana, but we'd have to travel by a shared taxi and that would take ten hours if the weather was fine, and up to twenty if there was flooding. By public bus, the journey would be longer: at least twenty hours and maybe up to thirty five if the roads were impassable. After considering the expense and the potential hardships, we decided on another Caribbean vacation. We so relished the easy way we'd vacationed in Tobago we wanted to have a similar experience again, but on a different island.

## Antigua
### (by Alastair)

So we started to look at which of the beautiful Caribbean Islands would be our next destination. There was an airline in Guyana called "Red Jet" offering unbelievable prices to certain Caribbean Islands, such as $49.99 plus taxes Cdn. round trip to Antigua. There were only a couple of seats at that price, but if you timed it right then the seats were yours. We jumped in and timed it right, and got the seats at that price.

On the internet, we found a charming, little cabin situated on a hillside in English Bay, just a short walk from the beach that seemed to have everything we were looking for, and it too was relatively cheap.

And we weren't disappointed. In fact, it far exceeded our wildest dreams in setting and beauty – we thought we were in paradise. Again! The main room of Stone Cottage, the little, stand-alone guest house, was mostly taken up by a huge king sized bed, tastefully dressed in pastel colored sheets and draped in a canopy of white netting. Two French doors opened onto the porch and served as a window to the breathtakingly beautiful grounds, resplendent with lush tropical bushes and brightly colored flowers. Four hummingbird feeders hung from the porch rafters, attracting hundreds of birds of different species and colors, but mostly the little yellow finches and gangs of humming birds. Our presence too attracted an array of small creatures, including lizards, which we fed by leaving banana and mango pieces on the patio railing.

By week's end, certain birds and reptiles had become so familiar with us they regarded us as their friends. One little guy, a yellow finch, got so friendly he regularly showed up at meal times and liked to partake of what we were eating. He jumped up on the rim of a plate and picked away, and sometimes alighted on the brim of a cup and bobbed for sips of coffee. He was out there in the morning when we arose, apparently looking for his morning coffee to start the day!

Our cabin was located about a half mile down a country road, which we ventured down in the mornings to catch a bus on the main road. Using local transportation (buses mostly) we found our way around the island and explored different beaches and towns, similarly to what we had done in Tobago.

Bill and Janine, the owners, also lived on the property but at a higher elevation on the hillside. Their view of town and the bay was stunning – they overlooked the harbor filled with water craft toys millionaires play with. To the side of their house was a large outdoor pool that we floated in every afternoon after we returned home from our daily outings.

Bill and Janine had lived in Antigua for over 30 years: she was from Britain and he was Canadian. She'd recently retired and had sold the two boutique shops she'd operated for over twenty years. He was still actively involved selling real estate and yachts. One night, they invited us to a gathering of expats to watch the movie "A Streetcar Named Desire" in a local art gallery, owned and operated by one of their friends. Most were retired and wealthy – they owned boats and properties – and had made Antigua their home many years ago. They were obviously enjoying a comfortable lifestyle on this beautiful Caribbean Island.

They say Antigua is the world's yachting capital, and a walk around the docks is an education on how the rich and famous live and spend their money. Many private yachts were so big we mistook them for cruise ships!

We were so comfortable with the cottage and grounds that most days we didn't want to go anywhere. We just lazed, read, wrote and played cards – pure luxury.

# Rupununi
### (by Alastair)

When my placement with YLA ended, I changed my status from active volunteer to "Accompanying Partner." This enabled me to accept a position with CESO, another Canadian volunteer sending agency, to do a four week placement in the Rupunini savannah, located between the Rupununi River and the borders of Brazil and Venezuela. What an incredible opportunity was that? I was now able to travel to this far corner of Guyana, and what was also fantastic, was that Candas would also get a chance to see this neck of the woods, or to be more precise the rain forest, because she was to join me for the last week after finishing her placement in Georgetown.

## Rewa Eco Lodge

All my life I'd been intrigued by documentaries on the Amazon rain forest, its rivers and wildlife, and the indigenous tribes who lived there. Never in my wildest dreams did I ever think one day I'd be there, working with the Amerindians, let alone being of service to them by sharing my skills and experience. So you can imagine how thrilled I was when CESO offered me a placement to go to four eco-lodges and spend four days at each in various parts of the Rupunini. Each Lodge manager had recently acquired a computer and someone at each lodge was learning to use it. My job was to review their manual accounting systems, develop processes to improve financial controls where needed, computerize the records by creating and inputting data into an Excel spreadsheet, and to train them on how to do it.

The Rupununi savannahs encompass 5,000 square miles of virtually untouched grasslands, swamps and rain-forested mountains, and are home to 15,000 Amerindians – the Wapisiana who live mainly in the south savannahs, the Makushi in the north, and some two hundred Wai-Wai who live in near isolation in the remote southeastern region bordering Brazil.

The savannahs are divided north from south by the Kanuku Mountains, Guyana's most biologically diverse region. The area supports a large percentage of Guyana's bio-diversity, including two hundred species of bird life, eighteen of which are native "only to the lowland forests of the Guianas." The Harpy Eagle, the world's most powerful bird of prey, an extremely rare and endangered species which once ranged the forests of South America, is now only found in the Rupununi/Kanuku mountain range.

The Rupununi is a paradise for eco-tourists. Designated as a "protected area" by the government of Guyana, it was home for over 80% of the mammals and 60% of the bird life to be found in Guyana's tropical forests and savannahs.

To get there, I flew in a small plane to Annai on the northern edge of the Rupununi. I could have gone by car, because the road from Georgetown to Lethem passed through Annai, but because I was on a schedule, Ceso didn't want to take any chances with the weather - it was unpredictable at that time of year. The road was originally a cattle trail before they upgraded it! Only about a hundred vehicles a week made the trip all the way, taking anywhere from twelve hours to two days to get to Lethe, depending on the weather.

Investors from Brazil, the region's rising power, want to pave the road and dredge a deep-water port near Georgetown to give northern Brazil a modern artery to export its goods to the Caribbean and North America. For Guyana, this idea holds both risk and reward – should it pursue global economic opportunities or a slower, more ecologically sustainable path toward development. This topic was hotly debated by all people living in the interior. Those in favor of paving it contend it would help alleviate Guyana's poverty, whereas others, especially the environmentalists, fear the loss of habitat, wildlife and even the traditional way of life for villagers. It would affect more than two million acres of rain forest.

I arrived in Annai with just a back-pack and a computer bag. From the airstrip, I walked to Rock View Lodge, two minutes away, where I was to meet Colin, the proprietor of Rock View, but he wasn't there, and neither was anyone else. What to do? I plumped down in a cozy chair, opened a book, and read. Colin eventually appeared – about an hour late – apologized with a half-smile for keeping me waiting, and then announced that Rovin, my Rewa guide, had been down at the river waiting for me since 7:00 a.m. It was now 1:30 p.m.!

We drove in Colin's jeep down to the river – about ten minutes away – where I met Rovin and boarded his aluminum boat with a high powered motor for the journey upstream to Rewa. The only way in and out was by river.

Rewa village is a small Amerindian community located in the north Rupununi, at the confluence of the Rewa and Rupununi rivers, and is home to about three hundred villagers from the Makushi tribe. The Rewa area is renowned for its abundance of wildlife and ecological diversity.

The Lodge, built in 2005 with a community grant from Conservation International, was co-managed by Dicky and Rudy – they worked alternate days. Catering staff of local villagers were called in when they had visitors. Conservation International is working with indigenous groups throughout the world helping them develop sustainable livelihoods whilst protecting the environment.

The main attractions at Rewa were the giant river otters and the endangered arapaima – the largest fresh water fish in the world, which grows up to ten feet long, weighs more than four hundred and forty pounds, and is known as the "dinosaur fish." But there were many other rare bird species naturalists from all over the world would like to see, including manakins, macaws, Guianan cock-of-the-rocks, hummingbirds, toucans and the spectacular harpy eagle. For the sport fisherman, there were tiger fish, piranha, arawana, payara and peacock bass to name a few. And for serious jungle trekkers, not me, for I'd only be there for four days and would be working most of the time, the chance to spot a jaguar, tapir, monkey, capybara, caiman, giant river turtle, puma, peccary, anaconda or agouti would be a dream come true. Because of its remoteness, isolation from humans, and the conservation efforts by the Rewa villagers, the Rewa River has the highest density of arapaima populations in Guyana, No human populations live upstream of the river's first twenty miles.

Two hours later, after a thrilling, fast ride up the river and its tributaries, cutting through dense rain forested land like a knife, past unnamed birds, whose screeching and cackling were new sounds to my ears, the tops of the thatched huts of the Rewa Eco Lodge came into view. We pulled over to the landing spot on the bank and a group of ladies hurried down the small hill to welcome me.

Exhilaration and curiosity filled my mind and body. I was living a dream. I felt more like a National Geographic photographer than an accountant when I stepped off the bobbing boat onto the wooden boardwalk and looked around the jungle clearing at the Lodge. Rudy, one of the co-managers, greeted me, showed me around and introduced his kitchen staff. They smiled at me, expressing their delight in having a visitor no doubt, for they now had a job for the day. Rudy's English was good, but no-one else spoke it.

The Lodge had two huts, each with two rooms containing two beds, and three separate self-contained cabins with two beds and an en-suite bathroom with a sink, flush toilet and shower. I had one of these. All beds had thick mattresses and mosquito nets. The large dinning benab served traditional meals of river fish and local vegetables and fruits. On a table, to one side of the eating area, were several bird and fish books I'd later peruse while waiting for meals to be served. And what was most amazing was they had internet, which only worked of course when they had power, which was from 6:00 p.m. to 9:00 p.m. every night when Rudy fired up the generator.

These eco-Lodges were a relatively new concept, and only a handful of visitors, mostly naturalists and researchers, had made the grueling trek into this one since it opened its doors. This was also the case for many of the Lodges I visited over the next four weeks.

Rudy took me on a leisurely stroll through the quiet village the following morning. The biggest structure was the single storied primary school. He did tell me how many students it had, but I've now forgotten – it wasn't many though for the whole village has less than three hundred inhabitants. As I ambled along, watching villagers periodically appear and disappear, I wondered what they were up to, what were they thinking about and how did they view the world. It dawned on me how substantially different my life was to theirs. It was as if I lived on another planet. I had mixed feelings about it. Ignorance is bliss so they say, and what you don't know you don't know. What a blessing that can be. These villagers looked blissful to me.

Late in the afternoon on the second day, Dillon took me out in a boat to see the giant river otters. There were about twenty in the water and on the bank opposite the Lodge and they were big, playful and curious. As Dillon raced the boat fast along the river's edge, the otters obviously took it as a sign we wanted to play, and swam alongside at full tilt. Where the land was cleared, the otters jumped out of the water, so fast that "flew" might be a more apt description, and ran along the bank matching us for speed. I'm sure they could've beaten the boat if they had a mind to, any time they wished. It was profound how child-like much of their behavior seemed to be – showboating and showing off their speed and maneuverability. Even their snorts sounded like human laughter, and I got a distinct sense they were laughing at me for being so ridiculous!

There was only one other visitor at the Lodge when I was there: a young American named James who was on a two month budget back-packing trip through South America. He was quite a hardy character who'd had an interesting journey over land and sea, and had many hair-raising stories to tell. He preferred to use his money on transportation and meals and not on luxuries, such as accommodations. He slept in a hammock hung from the rafters on the porch.

Over supper, which incidentally was prepared by the six kitchen staff as were all of my three meals a day for the next three days, I chatted with James about life, his travels and what he wanted to do when he got home. He was looking to find himself he said, but wasn't quite there yet. I gave him my copy of Echkart Tolle's, "A New Earth," hoping it might contain some of the answers he was looking for. He left at the break of dawn the next morning with Rovin.

I created Excel spread sheets to record sales, purchases, expenses and inventory and entered the data for the year from Rudy's paper records, and then held training sessions with him; Dicky, the other co-manager; and Sylvia, the administrative assistant. They beamed when they saw how the spread sheet captured the various expenses and how it instantly totaled the columns when you entered another number –so simple a program yet so powerful and useful in this setting. I used these spreadsheets as templates for the other Lodges I visited. It was so dark and clear at night that I saw the glory of the heavens in a splendored visual I'd never seen before. It was as if I'd taken the lid off the night sky and peered into what was within. The sky was full of sparkling objects and swirls, not empty as it usually appeared in most other places. I was gob smacked, and looked up most nights over the next four weeks to feast on a sight I knew I'd probably never see again.

Four days later, I packed up and boarded a small engine motor boat with Rovin. Some staff came down to the river's edge to wave goodbye. We headed out on the three hour ride to Guinea Landing, where I was to transfer to another boat for the last leg to Yupukari. To avoid the long bends in the river, Rovin took short cuts – ox bows – through mangrove infested channels and creeks, expertly maneuvering his boat through tangled roots and vegetation. On the way, I peered at the trees and water for birds and animals, but I didn't see any wildlife whatsoever. I wasn't disappointed. The forest was vast and the animals shy, and anyway, the experience itself was sufficiently rich and rewarding that it didn't matter.

Marcelus and Dillon, my guides on the last leg, were waiting for me at Guinea Landing when we arrived. I transferred my backpack and computer bag to their boat, thanked Rovin, and settled back for the two hour ride to Yupukari, a larger village than Rewa of the Makushi tribe. Dillon was in the back on steering duty while Marcellus up front acted as guide. It was too noisy for conversation so I just sat back and savored the visual grandeur of the river, the forested banks, and the unique experience I was having in the Amazon.

About three miles from our destination, we ran out of gas, and didn't have any extra cans in the boat. What to do? Dillon and Marcellus were beside themselves with laughter. I looked at them for a moment and then joined in the hilarity of the situation. Strange to say but I didn't feel the least bit uneasy. With a look and a smile that said it all, Dillon grabbed the only paddle in the boat and Marcelus picked up a flip flop sandal. I took the other one and we began paddling. We edged our way downstream, ever so slowly.

After we'd been at it for over an hour, we scooped our way over to the river bank where Dillon thought there was a dug-out canoe. It took a good half hour to do this because not only had we to battle the strong current, but we'd gone too far and had to double back to where the canoe was located.

I moved my personal effects into the canoe and we set off down a creek. The sunken gnarled roots and low hanging tree branches made progress almost impossible. Of course, I wasn't dressed for a trek into the bush and neither were my guides, but that didn't faze them. They knew what had to be done and set about doing it. Without saying a word, they rolled up their pants to the knees, jumped into the water, and Dillon put on my back-pack. They grabbed the canoe, one at the front and one at the back, and proceeded to manually steer, lift and pull it, with me in it of course, through and over obstacles. A half hour later, we reached land – well, a firm footing anyway – and set off down a path through hot, savannah type terrain to our destination at the Caiman House in Yupukari.

Marcellus said it was only two miles, but it sure felt more like four or five, and all under a hot sun with no fluids. How ill prepared was I for that scenario? What must I have been thinking? Or more precisely, not been thinking!

I stumbled into the Caiman House an hour later, exhausted and just a shadow of myself. Imagine Fernando's surprise – he was the manager – when his guest showed up three hours later than expected and on foot! I needed twenty minutes to sit quietly in a cool place with lots of water before I could coherently converse with him about my journey, and what I intended to do at the Caiman House. And one more thing – my lips were sun burnt - and I had trouble talking because my top and bottom flesh stuck together when I moved my lips. They were raw and sore and bothered me for weeks afterwards, despite the application of many salves and creams kind hearted people gave me. Everyone seemed to have just the right tube of something to "clear it up overnight," but they never did.

## Caiman House

Caiman House was originally built as a family home in 2003 by an American couple, Alice Layton-Taylor and Peter Taylor. Peter was a tropical ecologist and ex-Zoo keeper and wanted to conduct on-going studies on the Black Caiman. In 2005, with funding from the Saint Louis Zoo, Caiman House evolved into the Caiman House Field Station (CHFS), and Alice founded a U.S. non-profit organization called Rupununi Learners Foundation (RLF) with the objective of improving the reading abilities of Yupukari schoolchildren, and to financially support the Caiman project.

They hired and trained eighteen villagers on Caiman capture and data collection, and expanded the infrastructure to accommodate the increased activity. RLF financed the construction of a Guest House, and Wilderness Explorers, a Guyanan tourist operator, began to send paying guests. Boats took them out at night to watch the Caimans being caught, measured, weighed, sexed and tagged with a chip.

The construction wood shop became the solar-powered Yupukari Public Library (YPL), stocked with thousands of books and a dozen laptops connected to the Internet by satellite. Villagers were trained to manage the library. YPL began outreach to a dozen communities with book donations and bookshelf construction, resulting in a dramatic increase in the number of children qualifying to go to High School.

Caiman House was more accessible for tourists than the Rewa Lodge in that it was on a road, albeit a long, one hour dusty drive from Annai. Unlike deserted and sleepy Rewa, Caiman House was a happening place. Tourists, students and locals buzzed in and out, creating an energetic vibe that throbbed through the facility day and night. Many locals dropped in to see Fernando because he was so involved in Village affairs.

In addition to the main house with its sprawling rooms and gardens, there was a bank of accommodation units at the back for guests, similar to a motel in Canada. But Fernando treated me as a special guest and gave me the loft apartment above the main house. I took my leave early to have a good long sleep and recover from the physical hardship of the day's travel. It was hot, and because I was sleeping under a mosquito net, I felt comfortable lying in the nude.

Sometime during the night I awoke and sensed something light crawling up my back. I didn't panic. I quietly turned my head and looked down to see what it was. But I was too groggy to figure it out. I brushed my back with my hand throwing whatever it was at the mosquito net curtain, and then I lifted the curtain to drop whatever it was to the floor. It was fortunate there was a moon that night and its beams shone into the room because I could see what it was, and it was a bat! A small one, only about four or five inches long. It scurried over to the corner and tried to hide between my shoes. I turned over and went back to sleep. But sometime later, I awoke again sensing something was pulling on the mosquito netting. It was the bat – obviously trying to climb up to the top of the net. But why would it want to do that?

I found out the next morning at breakfast, when, after a good laugh about my bats in the belfry story, Fernando explained that bats cannot lift off the ground – they have to fall before they fly. He chuckled and added, "There's a fruit bat colony up there in the roof, but nothing to be alarmed about. They don't bite people. They eat mosquitoes. I thought they'd be good company for you."

I settled down to work, pulled up my work sheet template, customized it for the Caiman House, and entered data from their manual records. Fernando was computer literate, had a good mind for numbers, and asked if I could develop a "what if" spreadsheet to show the Lodges profitability at various guest volume levels. I relished this opportunity to do something above and beyond my stated objectives, and went at it with much gusto. I was both surprised and pleased with the result, knowing Fernando was grateful and as pleased as punch with his new financial planning tool. He kept inputting numbers and marveling at the financial outcomes and talking about them with his friend, Mike, a Canadian who assisted him in running the resort. Mike had been volunteering his time at Caiman House for many years, and from what he said, he intended to be there for many years to come. I felt relieved to have done something of real "value" because up until then I'd felt undeserving of the special treatment everyone was bestowing upon me in exchange for what I perceived to be so little.

On the second night, Fernando speared four small fish and pan fried them in butter for supper. I've forgotten what he said they were called, but they were delicious.

After supper, Dillon and Shamir took me out in a boat to "night spot" on the river: an attraction they offered tourists. We drove up river for about a mile, turned around, and quietly drifted with the current under a dark sky while shining powerful spot lamps into trees and water looking for eye-shines. It was a unique and wondrous experience. I saw many species of snakes of various sizes and colors wrapped around and hanging down from trees; night birds perched on branches patiently waiting for prey to fly or swim by; and I even saw a tree rat that spends its whole life in trees. And there were of course the black caiman, the largest member of the alligator family that can grow up to twenty feet long and weigh up to 2,400 lbs. Most adults in this river were about twelve to fifteen feet long and 500 lbs. Caimans are more agile and crocodile-like in their movements and have longer, sharper teeth than other alligator species.

Shamir paddled the boat to a river bank where there were many caiman eye-shines. Dillon picked up a baby, no more than eight or nine inches long, and gave it to me to hold. It was beautiful – so delicately complex and colored and perfect. Its eyes, legs and feet fascinated me. Once again, I was in awe of nature's genius. I so wished Candas could have been there with me to see and hold it.

Four days later, after doing the most I could in the time I had, I packed my bags, climbed onto the back of a motor bike and set off across the savannah to my next stop – the little village of Nappi and the Maipaima Eco-Lodge. Guy, my Maipaima contact, had set off before day break from Nappi with Ovid, another biker, to pick me up. I rode with Guy and Ovid took my back-pack and computer bag. We set out soon after breakfast, but apparently that wasn't early enough to see the giant ant eaters, armadillos and other creatures that lived in that strange land. The only wildlife I saw close up were giant egrets, but even so, they were quite the sight. They were definitely taller than any egret I'd previously seen.

It was an extraordinarily serene and pleasurable trek, albeit hot and uncomfortable. Sitting on a motorbike for two and a half hours, weaving around ruts on the road while hugging the driver lest I fall off, was a novel experience. The hot, flat grasslands, however, made a pleasant change from the dark rain forests and brown rivers that had been my environment for the last ten days. We were heading south under a bright blue sky with white puffy clouds towards the little village of Nappi, and in the distance, the Kanuku Mountain range shimmered on the horizon.

## Maipaima Eco Lodge

Nappi was a small, sprawling Makushi village located in the foothills of the Kanuku Mountains. The Maipaima Eco-Lodge, owned and operated by the community, was a gift from Foster Parrots Ltd., a non-profit organization dedicated to rescuing and providing sanctuary for unwanted and abused captive parrots and other exotic bird species at their New England Exotic Wildlife Sanctuary. Maipaima was located deep in the forest; about 10 km away from the village, and the only way to get in there was on a deeply rutted and muddy cow path.

Guy was not only my biker host, but he was also the Maipaima Lodge Manager and the mayor of Nappi and an election was coming up in two weeks. My timing couldn't have been worse. Guy was feverishly trying to make house calls to every constituent in his riding, which sprawled as far as the eye could see, and that was a long way. He didn't have time to spend with me on the project, and he didn't have anyone else to delegate the task to. If I had any questions, he suggested I ask Michael, my home stay host, and he'd take it from there!

I was surprised at this revelation because I assumed I'd be staying at the Eco-Lodge, as I did at Rewa and the Caiman House. They billeted me instead with Michael and Eleanor, two retired teachers, on their eclectic farm one kilometer down the road from the Nappi village. They had horses, cows, hogs, chickens, as well as a few cash crops.

My room was next to theirs in the main house. A separate building served as the kitchen. Here again, it was a most interesting opportunity to observe life through a different lens. Early every morning, I watched a vaquero (horse-mounted livestock herder) take the horses and cows off to pasture, and in the evening, just before supper, drive them back into the compounds for the night.

Guy came to the farm every morning for breakfast. He constantly apologized for not being able to meet with me during the day, asked if I had any unanswered questions, and enquired if I was being well looked after. I was. In fact, I was being treated as if I was royalty. Michael and Eleanor's cozy home, their interesting company and warm hospitality were more than I could ever wish for. Though they were busy with farm chores during the day, leaving me free to do whatever, we always sat down together at night for supper. We discussed a great variety of topics – politics, climate change, eco-tourism, village life etc., and had many laughs.

Michael said the government had given every villager a solar light. "Do you want to see them?" he asked. We went to the road, looked across the valley, and there, twinkling like bright stars in a night sky, were the solar lights. "They use them as Christmas lights to decorate their homes," he said with a chuckle.

It only took an hour the first day to prepare generic worksheets for the Lodge because I was using templates and didn't have any actual data to work with. There were three days to go before leaving for Lethem. What to do? I tried to rearrange the timing on the balance of my assignment, but I couldn't –everything was settled and seemingly cast in stone. The situation was beyond my control. I had to be pragmatic. I took a deep breath, changed my perspective, and was grateful. My new reality was that I was having a mini-farm holiday. Candas laughed her head off when I called and told her I was taking a few days off to holiday on a ranch in the Kanuku Mountains. And how fortunate that two other circumstances came into play, contributing to my enjoyment of the down-time: I'd packed my Kobo E-reader and had power, a rare commodity in those parts, to recharge the batteries. Michael had a generator and ran it every night from six to ten.

I went to Maipaima on the third day. Guy tried to rent Fernando's four wheel vehicle to take me in, but it wasn't available. He wanted me to see the Lodge and knew I only had one more day before I left for Lethem. So that's how I came to travel to Maipaima by ox cart with Felix and Oliver!

I will forever remember the trek for how could I possibly forget such a strange event. Early that morning after breakfast, I went to the road and waited in the middle of it for my transportation to arrive. I sensed that it would be worth the effort and I was right. I couldn't stop laughing when the wooden cart, pulled by two oxen, appeared on the horizon and lumbered down the hill towards me.

Felix and Oliver, the ox cart drivers and my guides for the day, were lighthearted fellows whose faces wore perpetual smiles – even when they weren't laughing. How could one not feel joyous in their company? And this was important because the ride itself was horrible. Every wrinkle in the hard, clay-surfaced road reverberated through the cart, and within a few minutes, the bones in my rear end were paining me. Thirty years ago, that might not have been a problem – there'd be more muscle and tissue on my bones – but at my age, it was an ordeal. Within minutes, I began wondering if I could make it through the day. We moved at a snail's pace.

It took at least thirty minutes to leave the village environs and pick up the cow path into the forest. There was some relief in that the ride wasn't as hard now on the back side, but the swaying of the cart, back and forth, front to back, as it climbed and tumbled over ridges, presented a new challenge – that of clinging on for dear life.

We had to abandon the cart twice to manually lift it out of a situation. The first was a mud hole that required sheer brute force to pull it out. The cart was in the mire up to its axle and the mud didn't want to give up its prey, but the oxen strained as we lifted, pulled and pushed, and we persevered and won out. I was on dry land of course because I was in dress shoes and dress pants. Felix and Oliver waded into the mud for the heavy lifting.

The second time was just before a small bridge over a stream. The left wheel slipped off the path into a rut tilting the wagon with all of us in it! But all wasn't disagreeable: glimpsing occasional clusters of green parrots and scarlet macaws perched on branches high in the canopy overhead was delightful; being startled by the piha's screaming was thrilling; and the cacophony of exotic bird songs and strange insect sounds ringing through the forest complimented the visual splendor and made the journey memorable.

As we slowly proceeded, yard by yard, the forest edge closed in on the path, shrinking the sky above and making our space darker and cooler. We'd entered the primary forest. And then suddenly, in a clearing, forty five minutes after we joined the cow path, the Maipaima Eco-Lodge appeared out of no-where. It was eerily quiet and uninhabited.

As Felix and Oliver unyoked the oxen to let them feed on the grass, I stood transfixed, taking in the astonishingly picturesque sight – more an illusion than a commercial enterprise. Three beautifully crafted wooden benabs, built in a traditional manner from local wood and thatch and constructed on stilts, were all connected by elevated, planked walkways. Much thought had obviously been given to keeping guests dry. The main benab housed the lounge/dining area, and the other two were guest cabins. Each contained four bedrooms with en-suite bathrooms and showers.

Why was no one there enjoying this Garden of Eden in Paradise? Probably it was because so few people knew about it, and secondly, because it was so off the beaten track, even for hardy adventurists. Such a pity! Maybe one day it will be more accessible, but they need to find a safer and more comfortable way to ferry guests in and out than the ox cart. I thought about returning with Candas once our placements ended - we could trek to the waterfalls, explore the bat cave, and search for the harpy eagle. Certainly something to consider, but even though we were physically located in Guyana, it would still be too expensive and time consuming for us pensioners on a budget.

Felix and Oliver bathed and reveled in the black water of a creek, washing off the mud and ooze from their legs while I hung out in a hammock and read. We topped off our visit by drinking the dozen beers my hosts had thoughtfully packed for the picnic!

The trip back was just as arduous, but quicker, as all homebound journeys seem to be. I was back at the farm in time for supper.

Early the next morning, I set off with Ovid on his motorbike for the two and half hour ride to Lethem, a town on the Guyana side of the Brazil border. I stayed there for two nights with a VSO couple who worked in Lethem.

My next stop was Bina Hill, back in the Rupununi, and to get there I took the public bus. It was a long, uncomfortable, but fortunately uneventful ride because it hadn't rained for a few days.

## Bina Hill

The Bina Hill Institute is primarily a secondary-level school teaching indigenous children core subjects of Maths and English, as well as natural resources management, forestry, wildlife management, agriculture, IT, and business and leadership skills. The Institute also serves as the major logistical hub for the region, and houses the North Rupununi Tourism offices and Board (NRTB) headquarters, and the broadcasting station and tower for the region's community radio service.

I spoke with Alphonso, the NRTB Manager, about my work at Rewa, Caiman house and Maipaima. He explained how the Board was working with and linking the Lodges to create diverse tourism packages for travel agents to sell to clients all over the world. Though in its infancy, I could see the huge potential in what they were doing and it was an exciting initiative.

The following day, I presented a work shop on how to build an Excel spread sheet to about twenty small entrepreneurs, involved in one aspect or another of the tourism and hospitality sector. They brought their computers, and though not all had Microsoft Office, we managed by sharing and saving to pen drives, and had a lot of laughs.

The following day, I set off for Kurupukari and the Iwokrama River Lodge: the last stop on my placement.

A driver from Iwokrama picked up three passengers from the Rock View Lodge in Annai and detoured to Bina Hill to get me. It was another long, bumpy ride, but by now I was used to it and didn't mind. Isn't it amazing how we humans can adapt with practice to anything?

## Iwokrama River Lodge

The Iwokrama International Centre (IIC) was established in 1996 under a joint mandate from the Government of Guyana and the Commonwealth Secretariat to manage the Iwokrama forest, a unique reserve of 371,000 hectares of rain forest, "in a manner leading to lasting ecological, economic and social benefits to the people of Guyana and to the world in general." The not-for-profit IIC organization was governed by an International Board of Trustees and managed by a professional team of around seventy permanent staff, located in Georgetown and Kurupukari. The Iwokrama Forest and Research Center (IFRC) provided a dedicated site in which to test the concept of a truly sustainable forest, where conservation, environmental balance and economic use could be mutually reinforced.

Drawing on its earlier work in sustainable forest management, the IIC was now in close collaboration with the Government of Guyana, the Commonwealth and other international partners, including the UK Company, Canopy Capital. They were developing a new approach to enable countries with rain forests to earn significant income from eco-system services and creative conservation practices.

As the IIC accounting staff in Georgetown did the books for the Center, my work at Iwokrama was to strengthen the inventory record keeping. As such, I created Excel worksheets, inputted manually prepared past data, and trained a number of clerks on how to do it.

As Candas would be joining me shortly, they gave me one of eight beautifully situated river facing cabins, each equipped with fans, bathroom, 24-hour electricity (supplied by solar power), and a wrap-around veranda with hammocks. It was idyllic, and I was eager for Candas to arrive. I was missing her terribly.

The grounds were spacious, manicured and contained two science laboratories, a huge air-conditioned conference center, and four banks of motel like rooms for visiting academics and researchers. It was only at meal times I became aware of how many other people were there. There must have been at least sixty or seventy, but it was hard to tell because people were coming and going at different times over the three hour supper time slot. The large, circular bar and restaurant overlooked the river, specialized in traditional Guyanese cuisine (river fish, rice and vegetables) and had the distinction of being the best food I'd tasted on my road trip. A sign on the wall in the restaurant for tourists made me smile. It read: "It's a privilege to see animals in their natural habitats: our experienced guides will take every measure to ensure your experience is enjoyable – the only thing you can kill is time; the only thing you can take is a photograph; and the only thing you can leave are footprints.

I went with the driver to Rock View to pick up Candas, who had flown to Annai from Georgetown. Though I had to work the next week, it would be a holiday for Candas: a chance for her to relax and savor a quiet, clean and refreshing environment after the many months she'd spent in noisy, polluted Georgetown. How fortunate were we to be able to spend our last week together in this tropical, scenic setting before returning to Canada. We must be living right for the Universe to keep sending us incredible, natural opportunities. Who needs to win a lottery when you have the Universe and Candas for partners?

To celebrate our year in Guyana, I wanted us to have one last memorable treat: an overnight at Turtle Mountain: Iwokrama Forest's signature vista. From Iwokrama, we would go by boat to the Turtle Mountain Base Camp located at the foot of Turtle Mountain near Paddle Rock Creek: a refuge for Brazilian tapirs, jaguars, peccaries, agoutis, and other wildlife. And from there, we'd hike the trail up the mountain as far as we wanted to go, all the while looking out for exotic birds, such as the harpy eagle, vultures, kites and hawks, and monkeys and other small animals. The views of the rain forest canopy from Turtle Mountain were said to be some of the most scenic in the world. We were so looking forward to this event, but it was not to be. It was too rainy and misty and the outing was canceled.

Were we disappointed? Yes, of course, but we'd learned to accept what was and didn't let it dampen our spirits. Just being at Iwokrama was sufficiently satisfying, and we treasured the chance to spend our last week together in Guyana in such a tropical setting.

Impulsively, and perhaps in response to having the Turtle Mountain trip fall through at the last minute, we decided to spend our last two overnights at Rock View Lodge before returning to Georgetown. And I'm glad we did. It gave us a chance to reflect on our most amazing past year and to mentally prepare ourselves for going home. Knowing that my work in Guyana was done felt like finishing a heavy tome. I'd closed the cover and was ready to consider what to read next. I felt satisfied and revitalized. I always relished the intermission period – the time between commitments – because possibilities and options excited me. And it was now doubly exciting because Candas was of this mind set too. We were two peas in the same pod.

*Rewa Eco Lodge*

*Rewa River*

*Fernando speared supper*

*Baby caiman*

*Caiman House Eco Lodge*

*Digging out the ox cart*

*Maipaima Eco Lodge*

*Iwokrama International Research Centre*

# Guyana
### continued (by Candas)

## Winding Down

As part of the monitoring and evaluation phase of the NVDP project NVD volunteers and staff facilitated focus groups and conducted surveys and interviews. As the reality of future funding shortfalls became all too apparent, we re-sized and modified next year's program activities, building on the program strengths and on areas where we'd had the most success in the past. With just three weeks left in my placement, I focused my attention on fundraising and developed a plan to seek stakeholders' support. To aid in this process I, along with another Cuso volunteer, facilitated a series of workshops for Guyana's Volunteer Leaders Network on Fundraising. Convincing the Guyanese Government to invest in this vital work would not only result in significantly improving the lives of persons with disabilities, but would boost citizen's pride of their nation.

I felt privileged to have been part of such a significant project and was gratified to have shared my skills and abilities to support my co-workers and volunteers. My placement came to an end and it was time to move on.

## The Journey Lies Within

Prior to leaving to meet up with Alastair for a week's vacation at the Iwokrama River Resort in the Rupinini, I began each day out on the back porch. I delighted in swinging on the hammock while watching and listening as the dawn gently birthed a new day. I meditated, shared my gratitude with the universe, and smiled with the thought of how the day would unfold.

In one way, I was sad about leaving because I'd grown and learned so much working alongside my co-workers and volunteers, whose strength of character, resilience and patience, in the face of great adversity, was so inspirational. I felt grateful for the opportunity to come to Guyana and offer my support to build the capacity of a national project, but most of all, I was truly thankful to the people whom I'd come to know and love.

## Last thoughts on Volunteering

Many of us feel a basic urge to help others, but we've not been able to do this because life's been too busy – raising a family, pursuing a career, catering to domestic and work obligations etc. With retirement come options and new opportunities. Whether we realize it or not, we all have a bank of skills, experience and talents that we've developed over our working lives we can use to help others, especially those in less fortunate situations. Being involved with and contributing to something outside our own life is part of the human way of being. Through experiences, we discover and learn about ourselves. Volunteering is about compassion, connection and the manifestation of a good heart wanting to engage with a cause or purpose far greater than ourselves. It isn't about receiving recognition, or anticipation of a return: it's about being in this world in ways that enhance all forms of life and create no harm. Volunteering provides us with life lessons which precipitate shifts in conscious living, personal growth and transformation. Each of us has a special gift to give to the world – ourselves. The path to serving others lies beneath our feet and our journey will lead us to living a purposeful and meaningful life.

GO FOR IT

# International Development Volunteering

The Canadian Government supports International Development through "Foreign Affairs, Trade and Development Canada. In 2014 the Canadian Government increased the number of countries of focus from 20 to 25, in order to deliver the greatest results for those in need. These countries were chosen based on their real needs, their capacity to benefit from development assistance, and their alignment with Canadian foreign policy priorities. The goal is to make Canada's international assistance more focused, more effective and more accountable.

Canadian organizations in partnership with Canadians, work to engage Canadian development expertise, interest and initiative. In order to achieve sustainable results, Canada works with hundreds of organizations around the world. Cuso International and CESO/SACO are two such Canadian organizations.

## CUSO INTERNATIONAL          www.cusointernational.org

Cuso International (Cuso) works to reduce poverty and inequality through the efforts of skilled volunteers. They place hundreds of volunteers of all ages, who collaborate with local groups, on projects in Latin America, the Caribbean, Africa and Asia. Volunteers share their expertise and perspectives, and unlock potential to create positive lasting change.

VISION STATEMENT:
A world where all people are able to realize their potential, develop their skills and participate fully in society.

MISSION STATEMENT:
Working in inclusive partnerships to overcome poverty through equitable and sustainable development.

## CESO/SACO          http://www.ceso-saco.com

CESO works to build stronger and more viable organizations and communities in both the private and public sectors in Canada and around the world. Their primary service areas are: strategic planning, business development, accounting and finance, organizational development, community development, governance and production and operations.

VISION STATEMENT:
We envision a world where there are sustainable economic and social opportunities for all.

MISSION STATEMENT:
We strengthen economic and social well-being in Canada and abroad through engagement of skilled and experienced Canadian volunteers working co-operatively with our partners and clients to create solutions that foster long-term economic growth and self-reliance.

# Epilogue

When we returned to Canada from Guyana, we decided to stay put for a while to enjoy the summer months in the company of our combined families (five children and seven grandchildren). As we were "homeless" – each having shed our homes and possessions prior to volunteering – and intended to be around for at least four months, we needed a place to stay that wouldn't inconvenience anyone. Once again the Universe intervened and brought to our attention a clean, furnished two-bedroom condo sublet in an excellent area and at a most reasonable price.

By summer's end, the urge to travel stirred in us again. This time, it would be to S. E. Asia. With only knapsacks and round trip air tickets in our hands – London, England to Bangkok, Thailand, we set off on a four month journey to explore the land, people and cultures of Bali, Vietnam, Laos, Thailand and Cambodia.

We intended to go to Thailand for the first month, but when we arrived in Bangkok it was raining hard, and still raining when we arose the next morning. Undaunted, we went to the airport, found an economically priced flight to Bali and went there instead. We seldom made advance hotel or travel reservations, preferring to stay uncommitted and thus available to respond to tips and information from other travelers and local people. Our experience was truly organic in nature and we just went with the flow, minute by minute.

After returning from Southeast Asia, we were again in need of a place to live. As a temporary measure, we volunteered to do a "Home and Pet Sit" assignment in Ottawa, Ontario for three months. The day before we left, a most surprising event happened that impacted our future living arrangements. Over breakfast that day, we discussed, for the first time, our need to get a place of our own that would be in keeping with our perspective of living minimally and leaving a small footprint on the earth. Within an hour of looking to see what was available, we found our idyllic future abode – a mobile home in a year round retirement community in central London. Two hours later, we met the real estate person, inspected the property and made an offer. By supper time, the deal was done with a closing date three months hence.

# Author Bios

### Alastair Henry

Disillusioned with the passivity of an early retirement and in search of greater personal fulfillment, Alastair sold up everything and went to live with a small First Nations band in a remote fly-in location in the N.W.T. The experience changed the direction of his life and he wrote about it in his memoir, "Awakening in the Northwest Territories."

When he left the north two years later, motivated and passionate about helping others, he went to Bangladesh as an International Development volunteer and then on to similar placements in Nigeria, Jamaica and Guyana.

### Candas Whitlock

After working in the Social Services Sector for over 40 years, Candas retired at the age of 63 and embarked on the most exciting and meaningful chapter of her life as an International Development volunteer.

*Today, Alastair and Candas are happily settled in London, Ontario, and busy themselves writing and staging Author Readings and Slide Show presentations at various venues throughout Ontario. Though their backpacks are now closed and put away, their minds are open and present, and should the Universe see fit to send them another interesting opportunity, you can be sure they will seriously consider it.*

*Alastair and Candas are co-writing another memoir about their budget backpacking adventures through Central America and Southeast Asia.*

# GO FOR IT